DISCOVERING
the
EARTH

OCEANS

A Scientific History of Oceans and Marine Life

Michael Allaby

Illustrations by Richard Garratt

Valparaiso Public Library
103 Jefferson Street
Valparaiso, IN 46383

Facts On File
An imprint of Infobase Publishing

OCEANS: A Scientific History of Oceans and Marine Life

Facts On File, Inc.
An imprint of Infobase Publishing
132 West 31st Street
New York NY 10001

Library of Congress Cataloging-in-Publication Data
Allaby, Michael.
    Oceans : a scientific history of oceans and marine life / Michael Allaby; illustrations by Richard Garratt.
        p. cm. — (Discovering the Earth)
    Includes bibliographical references and index.
    ISBN-13: 978-0-8160-6099-3
    ISBN-10: 0-8160-6099-1
    1. Ocean—Juvenile literature. 2. Oceanography—Juvenile literature. 3. Marine ecology—Juvenile literature. I. Garratt, Richard ill. II. Title.
    GC21.5.A45 2009
    551.46—dc22   2008033709

Facts On File books are available at special discounts when purchased in bulk quantities for businesses, associations, institutions, or sales promotions. Please call our Special Sales Department in New York at (212) 967-8800 or (800) 322-8755.

You can find Facts On File on the World Wide Web at http://www.factsonfile.com

Text design by Annie O'Donnell
Illustrations by Richard Garratt
Photo research by Tobi Zausner, Ph.D.

Printed in China

CP FOF 10 9 8 7 6 5 4 3 2 1

This book is printed on acid-free paper.

CONTENTS

Preface   ix

Acknowledgments   xi

Introduction   xii

CHAPTER 1
MAPPING THE OCEANS   1

Lucas Janszoon Waghenaer and His Mariner's Mirror   3

John Seller and His Atlas of the Ocean   5

Gerardus Mercator and His Map Projection   8

Maps: Drawing a Sphere on a Flat Surface   10

James Cook, the Greatest Chart-Maker   15

Robert FitzRoy, Surveying South America   21

FitzRoy and Darwin   26

Matthew Fontaine Maury, Ocean Currents,
and International Cooperation   29

CHAPTER 2
THE ORIGIN OF THE OCEANS   34

The Contracting Earth   35

James Dwight Dana and the Origin of Ocean Basins   37

Osmond Fisher and the Origin of the Pacific Basin   40

Conservation of Angular Momentum   41

CHAPTER 3
STUDYING THE OCEAN FLOOR   44

HMS *Challenger*   45

Charles Wyville Thomson, Scientific Leader of the
Challenger Expedition   48

Charles Bonnycastle and the Dream of Charting
the Ocean Floor   52

Reginald Fessenden and His Echo Sounder 54

Mid-Atlantic Ridge 58

William Maurice Ewing, Mapping the Ocean Floor 62

Mariana Trench 65

Harry Hess, Robert Dietz, and Seafloor Spreading 68

Tethys, Panthalassa, Pangaea, and the Drifting Continents 72

The Geologic Timescale 77

Plates, Ridges, and Trenches 79

CHAPTER 4
MEASURING THE DEPTH AND FLOW 85

Ferdinand Magellan, Measuring the Deep 85

El Niño 87

Lord Kelvin and How to Take Soundings from a Moving Ship 89

Antoine-Jérôme Balard and the Chemistry of the Oceans 93

Benjamin Franklin and the "River in the Ocean" 96

James Rennell, Who Mapped the Atlantic Currents 100

CHAPTER 5
JOURNEYS TO THE BOTTOM OF THE SEA 104

Diving Bells and Diving Suits 105

Pressure and Depth in the Ocean 106

Paul Bert and the Bends 108

Charles William Beebe and the Bathysphere 111

Auguste Piccard and the Bathyscaphe 115

The *Trieste* and Its Voyage to the Challenger Deep 119

Aluminaut, Alvin, and the Deep-Sea Submersibles 122

CHAPTER 6
LIFE IN THE ABYSS 127

HMS *Porcupine* and Life in the Porcupine Abyss 128

Galathea and the Philippine Trench 130

Inhabitants of the Dark Depths 133

Bioluminescence 136

Robert D. Ballard, Black Smokers, and Life at the Extremes 137

Extremophiles 142

CHAPTER 7
MONSTERS 145

Mermaid 147

Sea Monsters 150

Leviathan 152

Sea Serpents 155

Giant Squid 159

Oarfish, Sunfish, and Megamouth Shark 164

Sea Snakes 168

Coelacanth 170

Living Fossils 174

CHAPTER 8
MODERN EXPLORERS 176

Deep-Sea Drilling 177

Glomar Challenger 179

International Council for the Exploration of the Seas 180

Conclusion 184

Glossary 186

Further Resources 195

Index 199

PREFACE

Almost every day there are new stories about threats to the natural environment or actual damage to it, or about measures that have been taken to protect it. The news is not always bad. Areas of land are set aside for wildlife. New forests are planted. Steps are taken to reduce the pollution of air and water.

Behind all of these news stories are the scientists working to understand more about the natural world and through that understanding to protect it from avoidable harm. The scientists include botanists, zoologists, ecologists, geologists, volcanologists, seismologists, geomorphologists, meteorologists, climatologists, oceanographers, and many more. In their different ways all of them are environmental scientists.

The work of environmental scientists informs policy as well as providing news stories. There are bodies of local, national, and international legislation aimed at protecting the environment and agencies charged with developing and implementing that legislation. Environmental laws and regulations cover every activity that might affect the environment. Consequently, every company and every citizen needs to be aware of those rules that affect them.

There are very many books about the environment, environmental protection, and environmental science. Discovering the Earth is different—it is a multivolume set for high school students that tells the stories of how scientists arrived at their present level of understanding. In doing so, this set provides a background, a historical context, to the news reports. Inevitably the stories that the books tell are incomplete. It would be impossible to trace all of the events in the history of each branch of the environmental sciences and recount the lives of all the individual scientists who contributed to them. Instead the books provide a series of snapshots in the form of brief accounts of particular discoveries and of the people who made them. These stories explain the problem that had to be solved, the way it was approached, and, in some cases, the dead ends into which scientists were drawn.

There are seven books in the set that deal with the following topics:

- Earth sciences,
- atmosphere,
- oceans,
- ecology,
- animals,
- plants, and
- exploration.

These topics will be of interest to students of environmental studies, ecology, biology, geography, and geology. Students of the humanities may also enjoy them for the light they shed on the way the scientific aspect of Western culture has developed. The language is not technical, and the text demands no mathematical knowledge. Sidebars are used where necessary to explain a particular concept without interrupting the story. The books are suitable for all high school ages and above, and for people of all ages, students or not, who are interested in how scientists acquired their knowledge of the world about us—how they discovered the Earth.

Research scientists explore the unknown, so their work is like a voyage of discovery, an adventure with an uncertain outcome. The curiosity that drives scientists, the yearning for answers, for explanations of the world about us, is part of what we are. It is what makes us human.

This set will enrich the studies of the high school students for whom the books have been written. The Discovering the Earth series will help science students understand where and when ideas originate in ways that will add depth to their work, and for humanities students it will illuminate certain corners of history and culture they might otherwise overlook. These are worthy objectives, and the books have yet another: They aim to tell entertaining stories about real people and events.

—Michael Allaby
www.michaelallaby.com

ACKNOWLEDGMENTS

All of the line diagrams and maps in the Discovering the Earth set were drawn by my colleague and friend Richard Garratt. As always, Richard has transformed my very rough sketches into finished artwork of the highest quality, and I am very grateful to him.

When I first planned these books I prepared for each of them a "shopping list" of photographs, that I thought would illustrate them. Those lists were passed to another colleague and friend, Tobi Zausner, Ph.D., who found exactly the pictures I felt the books needed. Her hard work, enthusiasm, and understanding of what I was trying to do have enlivened and greatly improved all of the books. Again, I am deeply grateful.

Finally, I wish to thank my friends at Facts On File, who have read my text carefully and helped me improve it. I am especially grateful for the patience, good humor, and encouragement of my editor, Frank K. Darmstadt, who unfailingly conceals his exasperation when I am late, laughs at my jokes, and barely flinches when I announce I am off on vacation. At the very start Frank agreed this set of books would be useful. Without him they would not exist at all.

INTRODUCTION

Photographs of Earth taken from space show the planet to be blue. It is blue because that is the color of the oceans, which cover almost 71 percent of the surface. Despite their vast extent, however, until quite recently scientists were more familiar with the surface of the Moon than they were with the ocean floor. The Moon is visible from Earth, but the ocean floor is hidden in total darkness and comprises an environment that is extremely hostile to humans.

The oceans are also featureless. Travelers on land can orient themselves by recognizing landmarks and can navigate with the help of maps that show those landmarks. There are no landmarks at sea, so mariners must find other signposts to guide them. Many centuries ago navigators learned to steer by the Sun and stars and by compasses that align themselves with the Earth's magnetic field. Equipped with simple instruments, explorers crossed the oceans and surveyors mapped the lands that bound them. Then, having defined the boundaries of the ocean, scientists mapped the ocean currents. Finally, they turned their attention to the ocean floor and what lies beneath it. Scientists also speculated about how the oceans came to exist. What made the basins that the oceans fill?

Oceans, one volume in the Discovering the Earth set, tells of some of the mapmakers, scientists, and adventurers who dedicated their lives to improving our understanding of the oceans. The story begins with the long process of mapping the oceans and with the problems that had to be solved before this could be achieved satisfactorily—how, for instance, may the surface of a sphere be depicted accurately on the flat surface of a map? It tells of some of the greatest surveyors and mapmakers, including Captain James Cook (1728–79) and Robert FitzRoy (1805–65)—on whose ship, the *Beagle,* Charles Darwin (1809–82) sailed.

Having charted the coastlines surrounding the oceans, the story moves on to the exploration of the ocean basins. It describes some of the ideas about how these might have formed and how scientists approached the task of mapping the ocean floor. The book includes a brief account of the first, and to this day the most famous, of those explorations, made from 1872 to 1876 by scientists on board the adapted British naval ship *Challenger*. As they charted the deepest parts of the ocean, the scientists found evidence that explained the formation of ocean basins. This work led to the theories of seafloor spreading and continental drift, culminating in the modern theory of plate tectonics.

Eventually, their studies of the ocean made scientists eager to visit the ocean depths in person. People had been diving for oysters, and the pearls they sometimes contained, probably for at least 3,000 years, but even the most experienced pearl divers can remain below for only a very short time. The Romans used devices to allow longer dives, and the first diving suits date from early in the 18th century. The book describes some of those early diving suits and continues with the story of the first submersibles used for research and exploration, the bathysphere and bathyscaphe. These were followed by modern submersibles that have visited the very deepest places in the ocean.

Scientists believed the deep ocean floor was a desert, a place of perpetual darkness, intense cold, and huge pressure. When, at last, they were able to see the floor for themselves they found, to their surprise, that a wide variety of animals live there. Many of those animals were remarkable, but they were not the sea monsters of legend. The book describes some of those mythical monsters—sea serpents, giant octopuses, and mermaids—as well as real-life marine animals that are rarely seen and very large. The book ends with a brief account of the present state of ocean science and exploration.

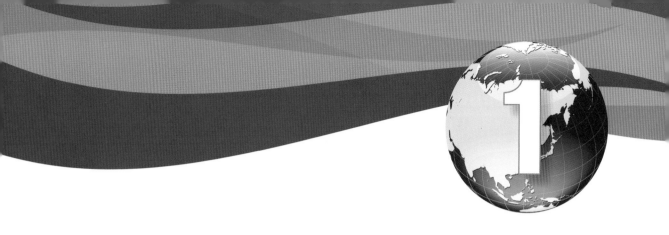

Mapping the Oceans

In April 2008 a team of archaeologists from the Swedish National Heritage Board discovered a hoard of about 470 coins not far from Arlanda Airport, just outside Stockholm. The coins date from the seventh to ninth centuries and were buried in about 850 C.E., which was a time when Norse mariners were plying the world, trading, plundering, and sometimes colonizing the lands they visited. It is not surprising that they acquired quantities of negotiable currency nor, in the days before there were banks and safe-deposit boxes, that someone should bury valuables and fail to return for them. What is special about these coins is that the inscriptions on them are in Arabic script. Some were minted in Baghdad, others in Damascus, and a few in Persia (modern Iran) and North Africa.

Norse ships were capable of long voyages, and their crews were highly competent. They could navigate out of sight of land. Some of them were Vikings—the name means pirates—who were greatly feared around the coasts of the British Isles, but the hoard of coins proves that the Norse mariners traveled much farther afield than Britain. They visited North African ports and Constantinople (now called Istanbul).

Nowadays ships' navigators use GPS (the Global Positioning System) to determine their position. Before the GPS satellites were made available, navigators used the direction of the noonday Sun to determine direction, and at night they steered by the stars, using compasses to maintain their heading. They understood tides and the locations and flows of ocean currents. Polynesian sailors navigate when out of sight of land by the appearance of the sea surface and sky.

Knowing the ship's position and heading is not enough by itself, however, because positions and headings are relative. If they are to mean anything useful, they must be plotted on a map. Polynesian sailors crossing between the islands of the South Pacific Ocean used to memorize maps. Young children spent years learning the maps from experienced sailors, who draw them in the dust on the ground. No doubt Norse sailors did the same, but all seafarers are liable to be carried off course by storms or feel the urge to explore new seas and new coasts, and in time the old ways proved inadequate. Sailors needed reliable maps.

This need presented a difficulty. It is one thing to map the overland route between two towns, using hills, rivers, and other natural features as landmarks. Coasts have headlands and inlets that can be depicted on charts and that a navigator can recognize. But how is it possible to map the ocean, far from land? Lewis Carroll (whose real name was Charles Lutwidge Dodgson, 1832–98) expressed it clearly in this following excerpt from his 1876 nonsense poem *The Hunting of the Snark: An Agony, in Eight Fits*:

> He had bought a large map representing the sea,
> Without the least vestige of land:
> And the crew were much pleased when they found it to be
> A map they could all understand.
>
> "What's the good of Mercator's North Poles and Equators,
> Tropics, Zones, and Meridian Lines?"
> So the Bellman would cry: and the crew would reply
> "They are merely conventional signs!
>
> "Other maps are such shapes, with their islands and capes!
> But we've got our brave Captain to thank"
> (So the crew would protest) "that he's bought *us* the best—
> A perfect and absolute blank!"

A map consisting of a blank sheet of paper might accurately represent the featureless expanse of the ocean, but it would be of little value as a navigational aid. Nevertheless, in time the problem was solved, and this chapter tells of the ways that were found to map the oceans. The story begins in the 16th century, when Europeans were transporting to Spain and Portugal the wealth they had seized in North America, which was then shipped to northern and western Europe. The Dutch controlled most of the trade around Europe, and the first reliable navigational charts were produced in the Netherlands.

The world is spherical. This presented the mapmakers with a further difficulty, since their maps had to be printed on flat sheets of paper. The chapter discusses how it is possible to draw a spherical surface on a flat surface and how Mercator developed his map projection, so comprehensively dismissed by the Bellman. It also describes the work of some of the naval *hydrographers* who explored and charted the oceans, one of whom was Robert FitzRoy (1805–65), who commanded HMS *Beagle* and was accompanied by Charles Darwin (1809–82).

LUCAS JANSZOON WAGHENAER AND HIS MARINER'S MIRROR

Until fairly recent times the North Sea contained vast shoals of herring, and prior to the 16th century North Sea fish, especially herring, made the Netherlands wealthy. It was said that Amsterdam was built on herring bones, and one visitor remarked that although there was not a single vineyard in all the Netherlands, nowhere was so much wine drunk. The country was also famous for its woolen and linen fabrics, yet its farmers neither raised sheep nor grew flax. Raw materials were imported, and until the 16th century everything was paid for with fish. In the 16th century the Dutch expanded their trading by importing, exporting, and transporting goods first around Europe and then around the world.

Spain and Portugal had claimed vast territories in the New World, and Spanish, Portuguese, and Italian ships regularly crossed the Atlantic carrying cargoes of valuable merchandise to ports in Spain and Portugal. Dutch traders then distributed these goods throughout northern and western Europe. To conduct this trade efficiently and safely, sea captains needed navigational aids.

The first of these appeared in the early 16th century and were called *routiers,* a French word that the Dutch and English corrupted to *ruttier* or *rutter.* Ruttiers contained written directions to guide mariners across European seas and along coasts. It may have been Portuguese sailors who began drawing sketch maps of coastlines in their logbooks. Sketches were obviously useful, and over the course of the 16th century they began to appear in Dutch ruttiers. Schools of chart makers had been established by the end of the century in the Dutch ports of Edam and Enkhuizen, to the north of Amsterdam, where cartographers prepared maps either from their own recollections and navigational experience or from descriptions they obtained from seafarers.

Edam and Enkhuizen are in the province of North Holland, and the chart makers became known as the North Holland School. They drew most of their maps on *vellum*, and many were elaborately decorated and prepared for clients who would hang them on their office walls like paintings. Others, less ornamented, were intended for use at sea.

Several distinguished cartographers belonged to the North Holland School, but one of the founders and most eminent was Lucas Janszoon Waghenaer (1533 or 1534–1606). Waghenaer was probably born in Enkhuizen, and he became a sailor, advancing to the position of pilot. Pilots board ships approaching port and guide them safely through the shallow water. Obviously, they know the coast and harbor approaches intimately. Waghenaer was at sea between about 1550 and 1579, during which time he gained a great deal of experience. In 1579 he settled down to life ashore, obtaining a post as a collector of harbor dues, but in 1582 he lost his job, perhaps because money went missing. From then on Waghenaer lived in comparative poverty, taking low-paid, temporary positions when he could find them and constantly trying to borrow money. All the time, however, he was working on a project that would revolutionize chart making. He was preparing an atlas containing charts as well as written directions. The first part was published in 1584, entitled *Spieghel der zeevaerdt* (Mariner's mirror). The second part appeared in 1585.

Waghenaer's "mirror" was an immediate success. Before long it had been translated into Latin, French, German, and English. Its 44 charts, all drawn to the same scale and style, showed European coasts from Gibraltar to Finland. Detailed sailing instructions accompanied each chart. Harbors and river estuaries were drawn to a larger scale. The work was of very high quality.

In 1592 Waghenaer produced a second book, *Thresoor der zee-vaerdt* (Treasure of Navigation) that also included instructions in astronomical navigation and described voyages to Asia. He published a third book in 1598, entitled *Enchuyser zeecaertboeck* (Enkhuizen sea-chart book). This time he included details of travels to Brazil, West Africa, and the Mediterranean. Lucas Waghenaer died in 1606 in Enkhuizen. He seems to have been living in poverty, because the municipal authorities had awarded him a pension, which they extended for a year after his death to help his widow.

JOHN SELLER AND HIS ATLAS OF THE OCEAN

The English were also a seafaring nation, but for exploration of the wider world they were obliged to use Dutch charts, for they produced few of their own. It was not until late in the 17th century that the situation began to change. In 1671 John Seller (ca. 1630–97) published the first volume of *Atlas Maritimus, or the Sea-Atlas: Being a Book of Maritime Charts,* a work he intended as a maritime atlas specifically for English mariners. On the title page Seller claimed it covered most of the known world, based on "the latest and best Discoveries that have been made by divers able and experienced Navigators of our ENGLISH NATION." He published the second volume and parts of the third and fourth volumes in 1672, but then he ceased working on the atlas and a few years later sold the publication rights.

Seller also published the first printed English map of the area around New York. This appeared some time between 1666 and 1674, entitled "A Mapp of New England." Seller derived it from English and Dutch sources but altered it and added detail in some places. One version of it appeared in his *Atlas Maritimus.*

The sale of publication rights on the *Atlas Maritimus* was forced on him. Seller was poor at business and in the end unsuccessful. In 1677 he managed to avoid bankruptcy only by entering into a partnership with a group of businessmen led by William Fisher, a printer, publisher, and retailer of books on navigation, and John Thornton, a chart maker. The partnership dissolved in 1679, leaving Fisher with the rights for the *Atlas Maritimus* as well as several other works by Seller and several plates from which maps were printed. Thornton took some of Seller's printing plates.

In an attempt to rescue his business, Seller began a collaboration with others in 1693 to produce an atlas of England and Wales, the *Atlas Anglicanus,* but had to abandon it with only six maps completed. He then embarked on a smaller atlas of the English and Welsh counties, the *Anglia Contracta.* This was more successful, and

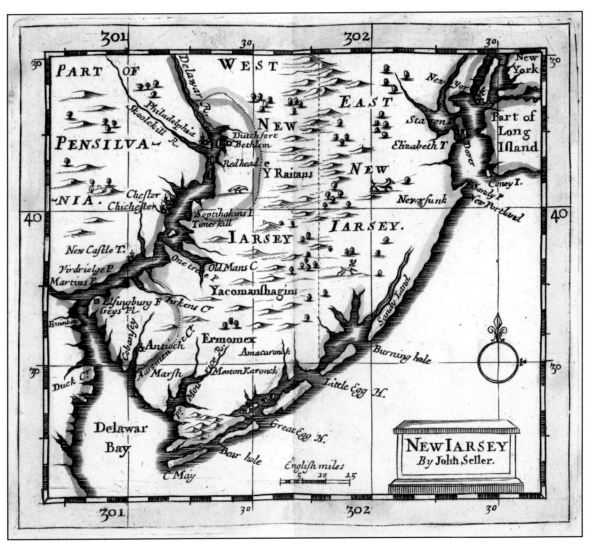

The map of New Iarsey (New Jersey), with parts of Pensilvania (Pennsylvania) and Long Island, drawn by John Seller (ca. 1630–97) and published in his 1682 *Atlas Maritimus.* The map shows Delaware Bay and the Delaware River extending northward to Philadelphia. This version of the map is taken from Seller's *A New System of Geography,* published in 1690. *(Darlington Memorial Library, Pittsburgh)*

several editions were published between 1696 and 1703. Other maps of North America, by Seller, Fisher, and Thornton, appeared in 1704, after Seller's death. The illustration shows Seller's map of New Jersey, which was published in 1690 in a book published by Seller and his son, John Seller, Jr., with the long title (in the original 17th-century spelling and punctuation): *A new systeme of geography, designed in a most plain and easie method, for the better understanding of that science. Accommodated with new mapps, of all the countries, regions, empires, monarchies, kingdoms, principalities, dukedoms, marquesates, dominions, estates, republiques, sovereignties, governments, seignories, provinces, and countries in the whole world. With geographic tables, explaining the divisions in each map.* In the same year Seller, Fisher, and Thornton published a map of the world, with charts of North America and the West Indies, and Seller published *Scripture Geography* containing maps of all the countries and places mentioned in the Old and New Testaments, *A book of Sea Stormes and Prospects,* the first part of *The English Pilot,* and, in collaboration with R. Adams, *A Book of the Prospects of the Remarkable Places in and about the City of London.*

John Seller augmented his income by offering lessons in "arithmetick, geometry, algebra, trigonometry, navigation, and gunnery; likewise the use of globes, and other mathematical instruments, the projection of the sphere, and other parts of the mathematics"—skills a ship's officer needed.

John Seller was also an instrument maker, surveyor, and artist. He had a shop called The Mariner's Compass in Wapping, not far from the Tower of London. Wapping in those days was a disreputable district of London, on the northern shore of the Thames, but for someone like Seller it was ideally situated. Wapping was where ships docked and sailors came ashore, bringing news of the distant places they had visited. Seller met them, talked to them, and based his maps on the detailed information he was able to glean from them. His ambition was to establish a chart-publishing business that could compete with the major Dutch publishing houses. In addition to his atlases, Seller published a star atlas, individual charts, coasting pilots—maps of the coast for the use of pilots—as well as almanacs, handbooks on navigation, and pocketbooks of maps. He also made and sold mathematical and navigational instruments.

John Seller had one advantage that came about in a curious way. He was a Baptist at a time when nonconformists were politically unpopular in England. Between 1649, following the public execution of King Charles I (1600–49), and 1660 England had been a republic; although Charles II (1630–85) was restored to the throne in 1660, it was only after a civil war. The causes of the civil wars—there were three in all—were complex, but religion played a significant part. Charles I sought to impose a uniform Church of England (Episcopalian) style of worship and prayer book throughout Great Britain. This provoked resistance from the Presbyterians in Scotland and Puritans in England. For years after the restoration of the monarchy the country remained unstable and the loyalty of anyone who did not belong to the Church of England was suspect. John Seller was tried and found guilty of conspiring to kill the king. This was a capital offense, but Seller was reprieved, possibly through the intercession of the king's brother, the duke of York (1633–1701), who later became James II of England (James VII of Scotland). In 1671 Seller was appointed Hydrographer to the King, a position he continued to hold during the reigns of James II (reigned 1685–88) and William III (1650–1702; reigned from 1689).

As Hydrographer to the King, John Seller was required to produce the best nautical charts possible, primarily for the use of the navy, but with this responsibility there came a commercial advantage. His position gave Seller a 30-year monopoly on the publication of such charts and atlases in England. Despite his poor business sense, Seller published so many maps, charts, atlases, and books that he succeeded in establishing a market for English-language maps and charts. Once the market existed, English cartographers were able to flourish.

GERARDUS MERCATOR AND HIS MAP PROJECTION

Throughout the Middle Ages mapmakers used surveying methods to draw maps and charts. They measured and recorded angles and distances, and they charted the oceans using compass bearings and distances that sailors estimated by the length of time they sailed on a particular heading and the speed at which they thought they were traveling. This is called *dead reckoning*, and although it works well enough as a rough-and-ready method of navigation, it is an unsatisfactory technique for compiling reliable maps. Surveying methods and instruments improved over the years, with resulting improvements in the quality of maps of land regions and countries, but there was a central problem to be resolved before maps of the world or charts of the oceans could be drawn accurately.

The problem was mathematical. Earth is approximately spherical in shape, and maps and charts are drawn on plane surfaces—they are flat. This presents little difficulty for maps of restricted areas such as a city or small country, because the distortion due to Earth's curvature is too small to be important. This is not true for maps of entire continents or oceans, or for maps of the world.

It is possible to draw a reasonably accurate map of the world onto a spherical surface, and in 1492, on a visit home, Martin Behaim (1459–1507) made for Nuremberg, his native city, what is now the world's oldest surviving globe. He called it the *Erdapfel* (earth apple). Behaim was a German navigator and geographer to King John II of Portugal. His globe is now housed in the Nuremberg National Museum, in Nuremberg, Germany.

Before setting sail, a ship's captain might use a globe to show the ship's owner the course to be followed, but the globe would be of no use at all for navigating the ship while at sea. Navigators need plane charts, and that was the problem. It is simple to demonstrate the impossibility of removing the surface of a sphere and laying it out flat. If someone peels an apple very carefully, removing all of the peel in one piece, the peel will still be spherical in shape, although it will readily collapse because it is no longer supported from inside. There is no way to lay it out flat without causing breaks and large distortions. A cartographer who wishes to transfer features from a spherical surface onto a plane map needs a method for achieving the transfer. That method is called a projection (see "Maps: Drawing a Sphere on a Flat Surface" on pages 10–11), and although it is simple in principle, there are several projections from which to choose, and all of them are highly mathematical in execution.

During the 16th century another school of Dutch cartographers formed in Leuven (Louvain), in the province of Brabant, which is now part of Belgium but was then Dutch. The school was founded by Regnier (or Renier) Gemma. Gemma was born in Dokkum, in the Dutch province of Friesland, on December 8 or 9, 1508; his parents were poor and both died while he was young. He suffered a physical

continues on page 12

MAPS: DRAWING A SPHERE ON A FLAT SURFACE

Earth has an *ellipsoidal* (also called *spheroidal*) shape, its diameter between the North and South Poles being shorter than its equatorial diameter. The difference is only 13 miles (21 km), however, and most small-scale maps assume that Earth is spherical. The distortions this assumption introduces become significant only with large-scale maps and maps that are required to show the land surface very accurately.

It is impossible to peel off the surface of a sphere and lay it flat. Consequently, mapmakers have had to devise techniques for transferring the features of a spherical surface onto a flat surface while minimizing the distortions this produces. The methods that are used are called *projections*. The operation is entirely mathematical. The mapmaker decides on a type of projection and then applies mathematical formulae to transfer points on the sphere to points on the plane surface of the map. It is easier to understand the choice of projection type if the mapmaker is imagined placing a flat sheet in contact with the sphere, although this is not what happens.

A flat sheet can be made into three shapes. It may remain flat, as a plane surface, or rolled to make a cylinder, or rolled to make a cone. Each of these shapes can be brought into contact with a sphere, and features from the sphere transferred to—projected onto—the flat sheet. There are, therefore, plane or *azimuthal* projections, conic projections, and cylindrical projections, with several versions of each. The illustration shows each of these. The azimuthal projection plots the surface features from a central point at which the plane touches—is tangent to—the sphere.

Any projection inevitably distorts the shapes, angles, areas, or distances of features on a spherical surface, but each projection has advantages for particular uses. The mapmaker must decide whether the map should show areas, distances, or angles and shapes correctly. It is impossible for any map to possess all of these properties. A projection that portrays areas accurately is known as an *equal-area projection* or *equivalent projection*. The disadvantage of this projection is that it badly distorts angles and shapes. An *equidistant projection* shows distances correctly, but only from certain points and in certain directions. A *conformal projection* shows angles and shapes accurately and the map scale is the same everywhere, so distances are accurate.

In a cylindrical projection lines of longitude *(meridians)* and latitude appear as straight lines that cross at right angles. The lines of longitude are evenly spaced, but lines of latitude are not spaced equally. All cylindrical projections show areas and shapes correctly only along a central line and along two lines parallel to and equidistant from the central line. The central line is usually the equator but may also be a central line of longitude. Shapes and areas are stretched—along an east-west axis if the equator is the central line—by an increasing amount with increasing distance from the central line.

A plane projection distorts the scale of features everywhere except where the plane touches the surface of the sphere. If the plane is tangent to the sphere, it touches at just one point and is then an azimuth projection. There

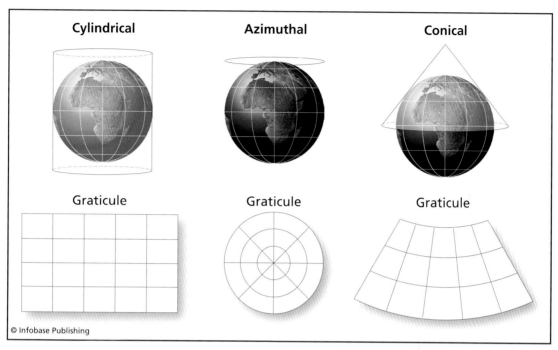

Cylindrical **Azimuthal** **Conical**

Graticule Graticule Graticule

© Infobase Publishing

Cylindrical, azimuthal, and conical map projections. In a cylindrical projection points on the spherical surface are projected to the places where they meet the surface of an imagined cylinder wrapped around the sphere. The azimuthal projection is a plane projection from a perspective directly above the North or South Pole. A conical projection is similar to a cylindrical projection, but the points are projected onto the surface of an imagined cone. The pattern of lines of latitude and longitude is called a graticule.

is no scale distortion at that point, but elsewhere the amount of distortion increases with distance from the central point. A secant plane cuts through the sphere, crossing the surface at two points. There is no distortion at those points, but distortion elsewhere increases with distance from the nearest point. In a plane projection lines of longitude and latitude appear as arcs or circles. Directions from the central point appear correctly.

A conic projection may be tangent or secant to the sphere. It distorts area and distance except along one line of latitude in the case of a tangent cone or two in the case of a secant cone.

Cylindrical projections are very widely used for nautical charts, the British National Grid system (used for surveying and in many maps), and for maps of the world. Maps in atlases, including road atlases, are drawn to cylindrical projections. Azimuthal projections are sometimes used to show distances along air routes and large areas of ocean. Conic projections are sometimes used in countries that are larger in an east-west direction than in a north-south direction.

continued from page 9

disability from which he partly recovered, but his health remained frail throughout his life. Following the death of his parents, his step-mother raised him. He attended school in Gröningen, and in 1526 he obtained a place reserved for poor students at the University of Leuven, where he gained a degree in medicine and stayed on at the university to study mathematics. Eventually, he became professor of medicine and mathematics at the university, and he also practiced as a physician in Leuven. When he became an academic, Gemma gave himself a Latinized version of his name, as was the custom. Having been born in Friesland, he became Regnier Gemma Frisius. Then he dropped his first name and was known simply as Gemma Frisius. He died in Leuven on May 25, 1555.

Gemma Frisius was his country's leading theoretical mathematician and had a particular interest in astronomy, geography, and mapmaking. In 1530 he used his mathematical skills to produce a globe that showed the world and also what he described as "the most important stars of the eighth celestial sphere." People could buy copies of his globe from workshops in Leuven, and Gemma Frisius wrote a book to accompany it, with a Latin title that in English is *On the Principles of Astronomy and Cosmography, with Instruction for the Use of Globes, and Information on the World and on Islands and Other Places Recently Discovered.* The book was published in Antwerp in three parts. In it, Gemma Frisius explains for the first time how to use a very accurate clock to determine longitude. Gemma Frisius also wrote many popular books on astronomy and geography.

In 1534 Gemma Frisius accepted a student of mathematics who was to become the most famous of all cartographers: Gerardus Mercator (1512–94). Mercator was the Latinized version of his name. He was born on March 5, 1512, in the hospice of St. Johann at Rupelmonde, in Flanders (now in Belgium), as Gheert Cremer (he was also called Gerard de Cremere). Gisbert, his father's brother, was a priest at the hospice. Gheert was the seventh child of Hubert Cremer, a farmworker and shoemaker, and his wife Emerantia. The family was poor and their life was hard. Often they ate only bread, because that was all the food they could afford. Despite the hardships, Gheert attended school in Rupelmonde, where he studied religion, arithmetic, and Latin. He could read and speak Latin fluently by the time he was seven years old.

Gheert might have aimed to become a priest like Gisbert, but the family finances deteriorated in the 1520s because of large tax increases that were needed to pay for a war against France, and in 1525 the Reformation started by Martin Luther (ca. 1483–1546) turned into a revolution. The anxiety and unremitting hard work undermined his father's health, and Hubert died in 1526 or 1527. Gisbert took Gheert under his wing and probably in 1527 sent him to be educated by the Brethren of the Common Life in 's-Hertogenbosch—despite the name, the order included sisters as well as brothers. The Brethren of the Common Life *(Broeders des gemeenen levens)*, founded in the 14th century, worked for their food and devoted themselves to education and literature. Emerantia died while Gheert was with the brothers and sisters, and it was then that Gerard chose his new name. *Cremer* means shopkeeper, and *mercator* is its approximate Latin equivalent, so he called himself, rather grandly, Gerardus Mercator de Rupelmonde.

The newly named Gerardus enrolled at the University of Leuven, graduating with a master's degree in 1532. The instruction he received was based largely on the writings of Aristotle. Mercator began to worry that Aristotle's account of the origin of the universe was different from the biblical account, and this undermined his confidence in the teachings of all philosophers. He decided not to work for a higher degree, left the university, and spent some time traveling. In 1534 he returned to Leuven and enrolled to study mathematics with Gemma Frisius. He also learned how to apply mathematics to geographic and astronomical problems.

At the same time Gaspard van der Heyden (1530–86) taught Mercator engraving and instrument making, and in 1535–36 Mercator engraved the copper plates that were used to print the paper strips from which Gemma Frisius and van der Heyden constructed a globe commissioned by the Holy Roman Emperor, Charles V (1500–58). It was the first time copper plates rather than wooden blocks had been used for printing, and they allowed for much finer detail. Mercator went on to make other globes, including one of the stars, and in 1537 he produced his first map, of Palestine. The illustration is of one of his maps drawn at around this time. It shows Turkey with parts of Romania and Bulgaria, many Greek islands, and Cyprus. It shows much detail and one mistake: Romania is really to the north of Bulgaria.

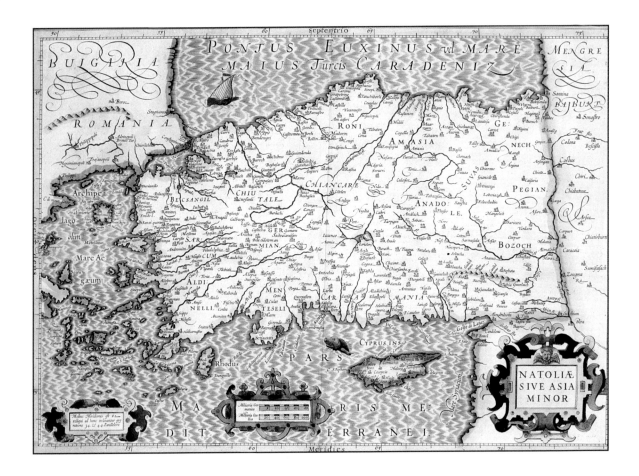

Natoliae Sive Asia Minor, a map of Turkey and the eastern Mediterranean that was rendered by Gerardus Mercator (1512–94) (University of Melbourne)

Mercator wished to produce a world map, which he would make by assembling separate maps of each region, but this proved difficult because of the vast amount of information, much of it conflicting and inaccurate, brought to him by sailors and other travelers. He realized, for example, that when a ship maintains a constant compass heading, in fact it follows a curved path called a *rhumb line* or *loxodrome*. Mercator traveled extensively to obtain reliable information on which to base his maps, and his travels contributed to the trouble in which he found himself when, in February 1544, he was arrested and charged with heresy. The charge related mainly to his Protestant beliefs, but someone who traveled widely was automatically suspect. He was imprisoned in the castle at Rupelmonde. Other suspected heretics arrested at the same time admitted they did not believe in certain parts of Catholic teaching and were executed by burning or being buried alive. The authorities searched Mercator's house but could find nothing incriminating; the University of Leuven supported him, and after seven months he was released.

In 1536 Mercator had married Barbara Schelleken, and the first of their six children was born in 1537. Many prisoners who were released were compelled to pay for the cost of their imprisonment, so Mercator probably received a large bill soon after returning home. Combined with the fact that while he was imprisoned he had been unable to complete work he had promised, this meant the family finances were gravely depleted. In 1552 the Mercator family moved to Duisburg, Germany, where a new university was planned, opening a profitable market for maps and books on navigation and geography. Mercator established a mapmaking workshop, and before long his reputation was secure as the leading mapmaker of the day. The family was also financially secure because in 1554 Mercator produced a map of Europe that sold well, and they moved into a large house in a fashionable neighborhood. From 1559 to 1562 he taught mathematics at a school where students prepared for university entry, but in 1562 the plans for a university were abandoned, and Mercator resigned from the school, his son taking his place.

In 1564 Mercator was appointed court cosmographer to Wilhelm, Duke of Jülich-Cleves-Berg (1516–92), nicknamed Wilhelm the Rich. It was during his time at court that Mercator developed the map projection with which his name is most often associated. He used it first in 1569 for a wall map of the world comprising 18 separate sheets. Mercator was also the first person to use the word *atlas* to describe a collection of maps in book form.

Gerardus Mercator suffered a stroke on May 5, 1590, which paralyzed his left side and made it impossible for him to continue working. He recovered slowly, and by 1592 he was able to work a little, but by then his eyesight was failing. He had a second stroke in 1593, which robbed him of the power of speech. He made a partial recovery and was able to speak a little, but then he had a third stroke, which was too much. He died in Duisburg on December 2, 1594.

JAMES COOK, THE GREATEST CHART-MAKER

As European nations expanded their trading networks, their need for reliable charts grew. Governments and merchants also needed information about new lands with resources that could be exploited and territories that could be claimed for strategic or commercial advantage. Scientific knowledge was also growing rapidly, and scientists needed information about remote parts of the world. Charting the oceans and surveying coastlines was time-consuming and costly, and voyages of exploration were often dangerous. The cost had to be met from public funds and, consequently, it was national navies that undertook the task. Naval surveying missions became increasingly scientific. James Cook (1728–79) was probably the first of the truly scientific explorers, navigators, and chart-makers. He was certainly one of the greatest.

James Cook was born on October 27, 1728, in the village of Marton-in-Cleveland, in North Yorkshire, England. Today the village is a suburb of the city of Middlesbrough. James was one of the five children of James Cook, a Scottish farm laborer, and his wife Grace, a local woman. Soon after James was born his father obtained a job as foreman on a farm at Great Ayton, on the edge of the North York Moors, owned by Thomas Scottowe, and that is where James spent his childhood. Scottowe paid for him to attend the village school.

In 1745 James was apprenticed to William Sanderson, a grocer and haberdasher in the fishing village of Staithes. He was strongly attracted to the sea, however, and Sanderson introduced him to John and Henry Walker, ship owners and coal shippers in the nearby port of Whitby. In July 1746 James's apprenticeship was transferred to them. They quickly recognized his ability and sent him to sea. In summer he would sail on colliers between ports along the eastern coast of England, and in winter he remained ashore, studying navigation and mathematics. His apprenticeship completed, James worked on ships trading with the Baltic Sea ports, returning to the Walkers in 1752 with the rank of mate. In 1755 the Walkers offered him his own command, but Britain was preparing for war with France—it would become the Seven Years' War, also called the French and Indian War—and James enlisted in the Royal Navy. On June 17 he began his naval service as an able seaman on HMS *Eagle,* a ship of 60 guns. Within a month he had been promoted to master's mate. By 1757 he had passed the examination qualifying him as a ship's *master* and was appointed master of the 64-gun HMS *Pembroke.* In the days of sail a ship's master was in charge of the ship's navigation and steering. He ordered the amount of sail to be carried, but he was not in overall command of the vessel.

In February 1758 the *Pembroke* crossed the Atlantic to take part in the siege of Louisburg, Nova Scotia, and the assault on Quebec, where James Cook played an important part in the charting of the St. Lawrence. After the battle he was transferred to HMS *Northumberland,* the flagship of Lord Colville. While serving as master of the *Northumberland,* Cook improved his knowledge of mathematics and astronomy and also learned surveying. He gained a reputation as a competent surveyor and in 1763, after a short spell ashore, he was appointed to the Newfoundland survey as master of the schooner *Grenville.* He spent five summers surveying the coast and the winters ashore working on his charts.

There was a solar eclipse in 1766, which Cook observed from the Burgeo Islands off the Newfoundland coast. His observations and his use of them to calculate his longitude were published in the *Philosophical Transactions of the Royal Society* for 1767, with a comment complimenting Cook on his mathematics. Cook had attracted the attention of the Royal Society of London and also of the Admiralty, and in 1767 he was commissioned as a lieutenant and given command of His Majesty's Bark *Endeavour,* a collier bought by the navy to carry observers to Tahiti, where they were to witness a transit of Venus across the Sun on behalf of the Royal Society. James Cook and the astronomer Charles Green (1735–71) were to be the two observers.

This was the first of what were to be Cook's three voyages of discovery in the Southern Hemisphere. The map shows the routes and dates of his three voyages. His orders were to ensure that the transit was observed, as the British contribution to an international effort to determine the distance between Earth and the Sun, and then to head to latitude 40° S in search of a supposed southern continent. If he failed to find it, Cook was to sail westward to New Zealand, then return home, rounding either Cape Horn or the Cape of Good Hope, whichever seemed best to him.

The *Endeavour* sailed from Plymouth on August 25, 1768, and dropped anchor at Tahiti on April 13, 1769. The observations of the transit apparently went well, but all observations of Venus were uncertain, and the calculations based on them varied so widely that the exercise was useless. Nevertheless, Cook established friendly relations with the Tahitians, and two of his passengers, the Swedish botanist

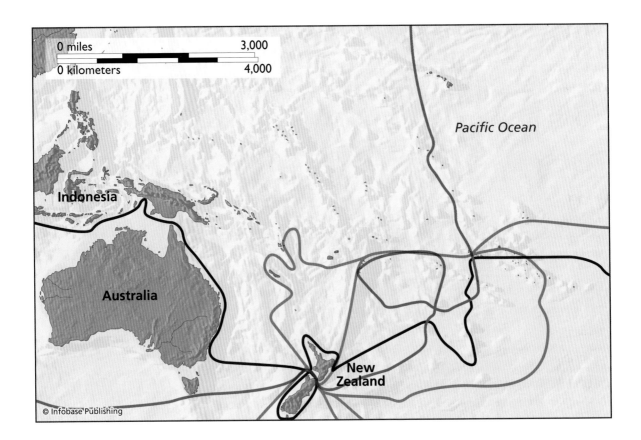

© Infobase Publishing

James Cook (1728–79) made three great voyages of discovery in the Southern Hemisphere. On this map his first voyage (1768–71) is shown in red, his second (1772–75) in green, and his third (1776–79) in blue.

Daniel Carlsson Solander (1733–82) and the English botanist Sir Joseph Banks (1743–1820), were able to collect many plant samples. The *Endeavour* left Tahiti on July 13, visiting and naming the Society Islands, the name referring to the way they are grouped together.

Cook failed to find the southern continent, so he headed for New Zealand, arriving on October 7. He sailed all the way around the New Zealand coast in a figure eight, mapping the entire coastline and making friends with the New Zealanders. Then he headed for Australia, arriving on April 19, 1770. Cook and members of his crew went ashore at the location of the present-day town of Kurnell on April 29, and Cook named that stretch of coast Botany Bay because of the many unique specimens the two botanists gathered. The *Endeavour* then headed northward, but the ship ran aground on the Great Barrier Reef and spent several weeks undergoing repair. They passed through the Torres Strait to Batavia (modern Jakarta, Indonesia), where malaria and dysentery broke out among the crew, with many fatalities. Cook then brought the ship home, sailing round the Cape of Good Hope and calling at Saint Helena, arriving back at Plymouth on July 12, 1771. He was immediately promoted to the rank of commander.

Determined to establish once and for all whether a southern continent existed, Cook proposed a second voyage and the Royal Society commissioned him to undertake it. His plan this time was to sail as far south as possible. He was given command of two ships, HMS *Resolution* and HMS *Adventure,* the latter commanded by Captain Tobias Furneaux (1735–81), and departed from England on July 13, 1772. He made three crossings of the Southern Ocean, crossing the Antarctic Circle on January 17, 1773, and reaching 71.17° S. He mapped South Georgia island and explored the South Sandwich Islands but failed to find the continent, although he remained convinced that it must exist somewhere to the south of the sea ice. The two ships parted company on October 30, 1773, off New Zealand when they were separated in fog. Furneaux took the *Adventure* back to Britain. Between February and October 1774 Cook sailed across the Pacific, visiting many smaller islands as well as Easter Island (Rapa Nui), the Marquesas Islands, the Society Islands, Niue, Tonga, Vanuatu, and New Caledonia. During this voyage Cook used a chronometer that was a copy of the one made by John Harrison (1693–1776). His logbooks were full of praise for this instrument, and the charts he produced were so accurate that some remained in use for almost two centuries.

The *Resolution* arrived back in England on July 30, 1775. Cook was promoted to the rank of post captain and given an honorary retirement, with an administrative position at the Navy's Greenwich Hospital. A post captain, addressed as "captain," was a naval rank as opposed to a courtesy title used of anyone commanding a ship. Once an officer reached the rank of post captain, further promotion was by seniority, and provided he remained in service a post captain would eventually attain the rank of admiral. Cook was elected a fellow of the Royal Society, which awarded him its Copley Medal for a paper he wrote on the methods he had used to combat scurvy. Famous and highly praised, Cook could have lived out the rest of his life in comfort, but soon he was planning a third voyage, this time to the north, to search for the Northwest Passage. This was a valuable prize, for the fabled route through the ice would allow ships to travel from the North Atlantic to the North Pacific without having to travel all the way around South America and brave the fearsome storms off Cape Horn. The route became much less important economically following the opening of the Panama Canal in 1914. The passage does open occasionally, but it was closed when Cook arrived. Roald Amundsen (1872–1928) was the first explorer to sail through it, in 1903. The passage was probably open in the early 1920s. It opened again in about 1940 and again, briefly, in 2007.

Cook set sail on July 12, 1776, once more in command of HMS *Resolution,* and headed south. HMS *Discovery,* commanded by Captain Charles Clerke (1741–79), joined him at the Cape of Good Hope. The two ships continued to Van Diemen's Land (Tasmania), New Zealand, Tonga, and the Society Islands before turning northward on December 8, 1777. On January 18, 1778, the expedition discovered the Hawaiian Islands, where they went ashore at Waimea Harbor, Kauai. Cook named the island group the Sandwich Islands, in honor of the Earl of Sandwich (1718–92), who was acting as First Lord of the Admiralty. The name fell into disuse during the 19th century.

After leaving Hawaii the ships headed northeast until they reached California, then explored and charted the western coast of North America as they sailed northward. They sailed around the Aleutian Islands and through the Bering Strait, reaching latitude 70.73° N before the pack ice became impenetrable, forcing them to turn back. Cook determined to spend the winter at Hawaii. After surveying part of the coast, the *Resolution* and *Discovery* anchored in Kealakekua Bay, Big Island, on January 17, 1779. They left Hawaii on February 4, but a topmast came free from its mounting, so Cook had to return to Kealakekua Bay for repairs. On February 14 some Hawaiians stole one of the *Resolution*'s small boats. This was a fairly common occurrence and dealt with by taking hostages and holding them until the stolen property was returned. Cook led a party ashore intending to capture a hostage, but the Hawaiians resisted and the men were forced to retreat to the beach. As Cook was helping launch the boats, the pursuing Hawaiians struck him on the back of the head, then stabbed him to death, and dragged his body away. Four of the marines in the party were also killed, and two were wounded.

Captain Clerke took command of the expedition and resumed the search for the Northwest Passage. He, too, was defeated by the ice to the north of the Bering Strait. Clerke died from tuberculosis, and an American sailor, Captain John Gore (1729 or 1730–90) took the expedition back to Britain, with Captain James King (1750–84) in command of the *Discovery.* They arrived home on October 14, 1780.

James Cook had married Elizabeth Batts (1742–1835) on December 21, 1762. They had six children, three of whom died in infancy. Two of his three surviving sons joined the navy, but all three were dead by 1794.

ROBERT FITZROY, SURVEYING SOUTH AMERICA

The seas around the British Isles are divided into areas for the purpose of weather forecasting. Forecasts and weather reports broadcast to ships refer to these areas. The map shows their arrangement and the location of coastal weather stations that monitor conditions in the sea areas. The sea area lying to the west of the Bay of Biscay used to be called Finisterre, but in 2002 this was one of several of the names that were changed. It is now called Fitzroy in honor of Admiral Robert FitzRoy FRS (1805–65), who in 1854 became Head of Meteorology at the Board of Trade, the precursor of the modern UK Meteorological Office. FitzRoy is remembered today as a meteorologist with a deep commitment to improving safety at sea. But he was much more than that.

FitzRoy was already a fellow of the Royal Society when he was appointed to head the emerging meteorological service. That honor was conferred on him in 1851 for his achievements in nautical surveying, scientific navigation, chronometric measurements, and for his published account of two surveying expeditions. Robert FitzRoy had surveyed the coast of South America. He was also a member of Parliament and the second governor of New Zealand.

Robert FitzRoy was an aristocrat. His father, General Lord Charles FitzRoy (1764–1829) was a direct descendant of Charles II (1630–85) and for a time served as an aide-de-camp to George III (1738–1820). His mother, Lady Frances Stewart (died 1810), was the eldest daughter of the marquess of Londonderry and the half-sister of Viscount Castlereagh (1769–1822), a senior politician who at different times was Secretary of State for War and Foreign Secretary. Robert was born on June 5, 1805, at Ampton Hall, in the village of Ampton, Suffolk.

Robert's formal education began in February 1818, when at the age of 12 he entered the Royal Naval College, Portsmouth. After graduating, on October 19, 1819, he entered the Royal Navy and was the first candidate ever to pass his final examinations with full marks. He was made a lieutenant on September 7, 1824, and served first on HMS *Thetis.* In 1828 he was appointed flag lieutenant (the naval

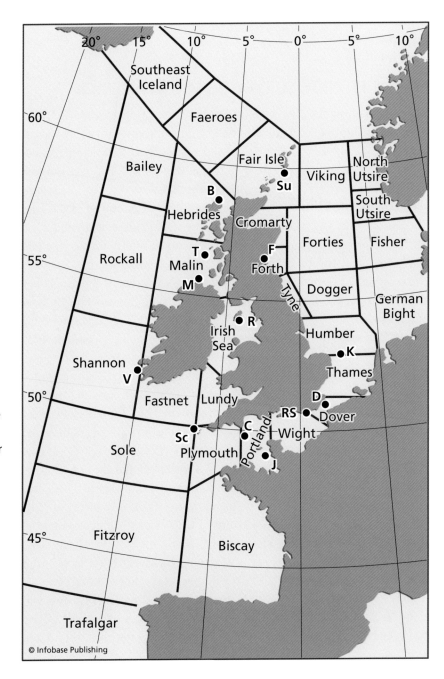

The map shows the sea areas around the British Isles that are used in maritime weather forecasts. The following letters mark the locations of coastal weather stations; forecasts begin with a report from each of these stations in turn: Tiree (T); Butt of Lewis (B); Sumburgh (Su); Fife Ness (F); Smith's Knoll Automatic (K); Dover (D); Royal Sovereign (RS); Jersey (J); Channel Light-Vessel Automatic (C); Scilly (Sc); Valentia (V); Ronaldsway (R); and Malin Head (M).

equivalent to an aide-de-camp) to Rear Admiral Sir Robert Waller Otway (1770–1846) on board HMS *Ganges.*

Otway was commander-in-chief of the South American station, where at that time two naval ships were surveying the coast of Patagonia and Tierra del Fuego under the overall command of Captain Phillip Parker King (1793–1856) on board HMS *Adventure.* The second ship, HMS *Beagle,* was working in the waters around Tierra del Fuego when her captain, Pringle Stokes, fell victim to severe depression. When the ship reached Port Famine on the Strait of Magellan, Stokes remained locked in his cabin for 14 days. When he finally emerged he went ashore to a remote spot where on August 2, 1828, he shot himself, dying 12 days later. Captain King appointed Lieutenant W. G. Skyring to take the *Beagle* to Rio de Janeiro, where it arrived on December 15. Admiral Otway appointed Robert FitzRoy, then 23 years old, as temporary captain of the *Beagle.* FitzRoy completed the survey and arrived back in England on October 14, 1830.

While members of the *Beagle* crew were ashore in Tierra del Fuego, a group of local people stole their boat. The *Beagle* set off in pursuit and finally captured the thieves' families, bringing four of them on board to be held as hostages until the boat was returned. When it proved impossible to set the captives ashore, FitzRoy decided to teach them English and the rudiments of Christianity. Eventually he took all four back to England with him, hoping they might prove useful as interpreters and also that exposing them to life in England might make them grow friendly toward English people. The *Beagle's* crew gave the four names: York Minster (after a large rock near where he was caught), Jemmy Button (because FitzRoy paid his family a single mother-of-pearl button for him—it is not certain whether they parted with him willingly), Fuegia Basket (because the boat the Fuegians returned in place of the ship's boat resembled a basket), and Boat Memory. Newspapers published stories about them, and they became famous. They were presented to the king, William IV, and Queen Adelaide presented Fuegia Basket with a bonnet. Unfortunately Boat Memory died after being given a smallpox vaccination. A trainee missionary, Richard Matthews, cared for the others.

Robert FitzRoy stood for election to Parliament in 1831 but was defeated. He was anxious to return the three Fuegians to their home and had decided to charter a ship for this purpose at his own expense, when Captain Francis Beaufort (1774–1857), head of the

Hydrographic Office of the Admiralty, and FitzRoy's uncle, the Duke of Grafton (1760–1844), interceded with the Admiralty on his behalf. The Admiralty was planning a second survey expedition to South America, using HMS *Chanticleer,* a 10-gun brig, but the ship was in poor condition and the *Beagle* was ordered to take her place. On June 25, 1831, Robert FitzRoy was appointed to command her.

The *Beagle* needed considerable work to make her suitable for the task she had been given, and in July she was taken to Devonport, Plymouth, to be refitted and partly rebuilt. FitzRoy had the upper deck raised. This made the ship more stable and helped water drain faster from the decks in bad weather. He had additional sheathing added to the hull, and he had her fitted out with the very latest chronometers and other instruments—for which he paid out of his own pocket. The illustration shows the refitted *Beagle* off the South American coast.

HMS *Beagle* off the South American coast. The *Beagle* was a six-gun bark (or barque), launched in 1820 and finally broken up in 1870. She was 90.3 feet (27.5 m) long and 24.5 feet (7.5 m) across the beam, and carried a crew of about 70. *(Science Photo Library)*

FitzRoy was well aware of the loneliness of command and of the contribution it made to the tragic death of Pringle Stokes. In the early 18th century it was impossible for a person of FitzRoy's rank to mix socially with his crew, and on this voyage the *Beagle* would have no companion ship with its own captain. It meant that for the five years the *Beagle* would be away from home he would have no company at mealtimes or during his leisure hours, and no one with whom he could hold conversations. He invited a friend to accompany him, and when this failed he turned to Captain Beaufort for advice. He needed someone who shared his scientific interests, would seize the opportunities the voyage offered to study the plants, animals, and geology of the places they visited, and who would be his companion. Through his contacts, Beaufort approached the naturalist Leonard Jenyns (1800–93) and the English botanist and geologist John Stevens Henslow (1796–1861), Regius Professor of Botany at the University of Cambridge. Both turned him down, but both recommended the 22-year-old Charles Darwin (1809–82). FitzRoy and Darwin spent a week together getting to know each other, and finally FitzRoy accepted Darwin. Darwin had to pay his own expenses and provide such scientific equipment as he thought he would need. Charles Lyell (1797–1875), the professor of geology at King's College, London, had asked FitzRoy to record his observations of geological features, and before they sailed FitzRoy gave Darwin a copy of the first volume of Lyell's *Principles of Geology,* published in 1830, to help him prepare. The following sidebar describes the subsequent relationship between the two men.

After several delays, the *Beagle* sailed from Plymouth on December 27, 1831. She returned to England almost five years later, arriving at Falmouth, Cornwall, on October 2, 1836. The map shows the route of her round-the-world voyage, calling at the Cape Verde Islands, following the South American coast, passing through the Strait of Magellan, and visiting the Galápagos Islands, Tahiti, New Zealand, Australia, the Maldives, and Mauritius. FitzRoy completed the surveying mission, sailing through some of the world's most dangerous waters, with no damage to the ship and the loss of only six lives—one of whom, the purser, died from old age.

Soon after returning to England FitzRoy married Mary Henrietta O'Brien (1812–52), then settled down to write an account of the voyage. This appeared in 1839 entitled *Narrative of the Surveying Voyages*

FITZROY AND DARWIN

Robert FitzRoy was anxious to find a companion for the five-year, round-the-world voyage of the *Beagle*, and he was very clear about the kind of man he was seeking. FitzRoy was keenly interested in science and natural history and stipulated that the successful applicant must be a trained naturalist. In his letter to Captain Beaufort he asked for "some well-educated and scientific person." Obviously, he must be intelligent, so the two could converse at an intellectual level FitzRoy found congenial. Finally, the traveling companion must be a gentleman of high social standing, so they could dine together. FitzRoy was an aristocrat of the highest English social class, and in 19th-century England people from widely different classes did not mix socially. It would have been unthinkable for Robert FitzRoy to take his meals at the same table as someone from a lower class. Charles Darwin did not belong to a titled family, but his grandfather Erasmus Darwin (1731–1802) had a considerable reputation as a physician and poet, and Charles's mother was a daughter of Josiah Wedgwood (1730–95), the owner of the flourishing pottery business. Darwin had to persuade the captain of his social acceptability and when they first met FitzRoy, who believed he could judge character by the shape of a person's face, came close to rejecting Darwin because of the shape of his nose. Nevertheless, the two got along well and soon became firm friends. Their friendship lasted for many years after the voyage had ended.

Accommodation on board the *Beagle* was cramped. The ship had three officers, as well as Richard Matthews, a missionary accompanying the three Fuegians FitzRoy was taking back to Tierra del Fuego. Darwin lived in the chart room in the stern of the ship, one deck above Captain FitzRoy's cabin. The chart room measured nine feet (2.7 m) by 11 feet (3.3 m) and was five feet (1.5 m) high. The mizzenmast rose through the room, and the large chart table occupied most of the center. Darwin slept in a hammock slung above the chart table.

of His Majesty's Ships Adventure *and* Beagle *Between the Years 1826 and 1836, Describing Their Examination of the Southern Shores of South America, and the* Beagle's *Circumnavigation of the Globe.* Two volumes of this work, plus a separate appendix, described the *Beagle*'s two voyages. Darwin contributed a fourth volume, *Journal and Remarks,* which was later published by itself as *The Voyage of the* Beagle. In 1837 Robert FitzRoy was awarded the gold medal of the Royal Geographical Society, and in 1851 he was elected a fellow of the Royal Society, sponsored by 13 existing fellows, including Charles Darwin and Francis Beaufort.

With the round-the-world voyage completed, Robert FitzRoy was well placed for a career on land. He was elected Member of

Darwin was not the official naturalist on the voyage. Robert McCormick, the ship's senior surgeon, held that position. But FitzRoy favored Darwin, affording him the best opportunities to visit interesting sites and to collect specimens. As a companion, Darwin paid about £500 for his food and accommodation—about £36,000 ($72,000) in today's money—so his position was more like that of a paying passenger. It meant that he was permitted to keep all the specimens he collected; specimens collected by an official naturalist were government property, although he might be allowed to keep some. In addition, FitzRoy agreed in advance that Darwin would be allowed to spend as much time ashore as he wished. On several occasions Darwin went ashore, traveled overland, and rejoined the ship at a later port of call.

Robert FitzRoy had a fiery temper—his nickname was "Hot Coffee"—and quarreled with Darwin on several occasions but was always quick to apologize and restore good relations.

Their disagreements were mainly political. FitzRoy held deeply conservative views; Darwin was more liberal. Disagreements over religion began to emerge during the voyage and intensified after the 1856 publication of *On the Origin of Species*. At first their religious views were very similar, but his observations increasingly led Darwin to question the fixity of species, while FitzRoy struggled to reconcile the great age of the Earth described in Charles Lyell's *Principles of Geology* with the biblical account. FitzRoy later wrote of having read "geologists who contradict, by implication, if not in plain terms, the authenticity of the Scriptures," and that he had once remarked that a large plain of sedimentary material could not have been deposited by a 40-day flood, then criticizing himself for doubting scripture. He then argued that seashells found in rocks high in the mountains are proof of Noah's Flood. This divergence finally caused a split between them, but for years after the *Beagle* returned to England the two remained close friends.

Parliament for Durham in 1841 and worked to improve conditions for merchant sailors and fishermen. He was also appointed a conservator of the River Mersey. The Mersey carried shipping to the important port of Liverpool, and FitzRoy supervised a survey of the condition of the river. Then, in 1843, he was appointed governor of New Zealand. This was not a success. He objected to the way British settlers treated the Maori people, and when there was a serious dispute over land rights, FitzRoy took the Maori side. In 1845 he was recalled to London and naval duties. In 1848 he oversaw the fitting out of a new *frigate*, HMS *Arrogant*. FitzRoy recognized that steamships would soon replace sailing ships. *Arrogant* was a sailing ship that also had a steam engine, and FitzRoy was made its captain.

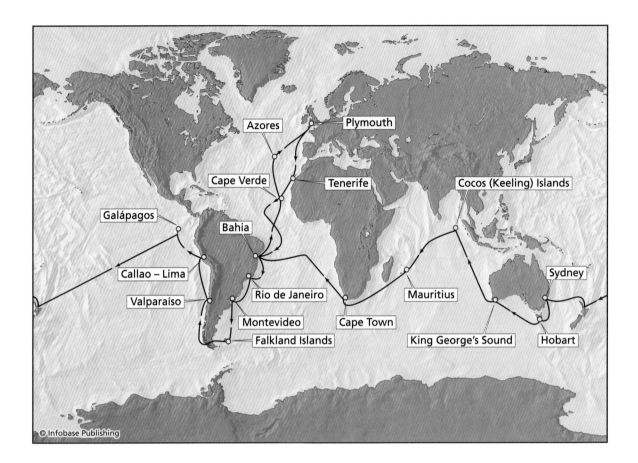

The route followed by HMS *Beagle*. She sailed from Plymouth on December 27, 1831, and arrived at Falmouth, Cornwall, on October 2, 1836.

In 1854, on the recommendation of the president of the Royal Society, John Wrottesley (1798–1857), FitzRoy was appointed meteorological statist to the Board of Trade. *Statist* was the 19th-century name for a statistician, and the Board of Trade was a government department, equivalent to the U.S. Department of Commerce. With a staff of three, FitzRoy established what later became the Meteorological Office.

Robert FitzRoy was working hard, his office issuing regular weather forecasts, and he enjoyed family life. He and Mary had two daughters and one son. Mary died in 1852, and in 1854 FitzRoy married Maria Isabella Smyth (died 1889); they had one daughter. By the 1860s he had risen to the rank of vice-admiral. But he had always been a troubled man. Criticism wounded him deeply, and newspapers mocked his weather forecasts when these proved wrong. The distinguished American naval scientist Matthew Maury (1806–73; see "Matthew Fontaine Maury, Ocean Currents, and International Cooperation," below) criticized his scientific approach to meteorology. Most of all, however, FitzRoy worried about the part he had played in the development of Darwin's evolutionary theory, with which he disagreed profoundly. All of these concerns finally overwhelmed him. He fell victim to a deep depression, and on April 30, 1865, at his home in Upper Norwood, near London, he took his own life. In his autobiography Darwin wrote of him: "FitzRoy's character was a very singular one, with many noble features; he was devoted to his duty, generous to a fault, bold, determined, indomitably energetic, and an ardent friend to all under his sway."

MATTHEW FONTAINE MAURY, OCEAN CURRENTS, AND INTERNATIONAL COOPERATION

The first ship of the U.S. Navy to circumnavigate the world was the *sloop of war Vincennes,* known at the time as "the fastest sailer in the Navy." A sloop of war was a small three-masted warship with a single gun deck, equipped with up to 18 cannon. In 1829 the *Vincennes* had been stationed in the Pacific when she was ordered to return home by sailing westward around the world. One of the *midshipmen* on board the *Vincennes* was Matthew Fontaine Maury (1806–73). Maury was later to be known as the "Pathfinder of the Seas" for showing sailors how to use the winds and ocean currents to shorten long sea journeys. He devised a system to standardize the recording of winds and currents that was adopted throughout the world, and he charted the migration routes followed by whales, which greatly increased the size of the whale harvest. He was also known as the father of modern oceanography and naval meteorology and as "scientist of the seas."

In *The Hunting of the Snark* the ship's crew may have been impressed by a map of the ocean that showed nothing at all, but in reality the ocean is not featureless: There are ocean currents. The map shows the principal ones. Currents flow like rivers through the ocean, and ships can take advantage of them—provided navigators know their location—sometimes with spectacular results. For instance, in 1848 the merchant bark *W. H. D. C. Wright* carried a cargo of flour from Baltimore to Rio de Janeiro and returned to Baltimore with a cargo of coffee, taking 75 days for the round-trip—a full month faster than the schedule had allowed. The ship made such good time because the captain navigated with the help of two pamphlets by Matthew Maury: *Wind and Current Charts* and *Sailing Directions,* both published in 1847. Matthew Maury had charted the currents and the prevailing winds. Maury's work also speeded the ship's passage by showing a route through the doldrums.

In the days of sail the doldrums were regions of the tropical oceans that sailors dreaded. They were places where winds were light, or the air was still, and ships could remain becalmed for so long that their supplies of drinking water were exhausted. Samuel Taylor Coleridge (1772–1834) described the plight of sailors becalmed in the doldrums in the following excerpt from *The Rime of the Ancient Mariner:*

All in a hot and copper sky,
The bloody Sun, at noon,

The map shows the principal ocean currents: Currents in the North and South Pacific, North and South Atlantic, and Indian Oceans follow approximately circular paths called gyres.

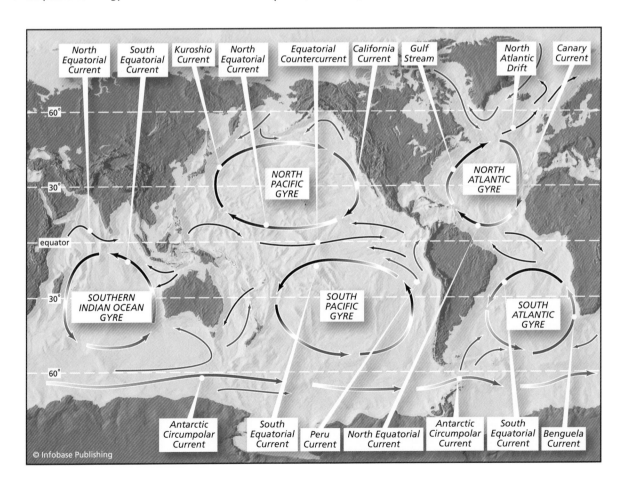

Right up above the mast did stand,
No bigger than the Moon.

Day after day, day after day,
We stuck, nor breath nor motion;
As idle as a painted ship
Upon a painted ocean.

Water, water everywhere,
And all the boards did shrink;
Water, water everywhere,
Nor any drop to drink.

Maury discovered that the doldrums occur in particular places and there are ways to navigate between them. He compared them with mountains on land that present formidable barriers to the traveler but that have passes and gaps between them. By showing navigators a way through the doldrums, Maury must have saved many lives.

Matthew Fontaine Maury was born on January 24, 1806, in Spott-sylvania County, Pennsylvania. His grandfather, the Reverend James Maury (1719–69), was an Episcopalian minister who also ran a school in Albemarle County, Virginia, where future presidents Thomas Jefferson, James Madison, and James Monroe were pupils. When Matthew was five years old the family moved to Franklin, Tennessee. Matthew's elder brother, John Minor Maury (1795–1828), was a naval officer, and Matthew wanted to follow in his footsteps. Unfortunately, John died from yellow fever, caught while chasing pirates, and their father forbade Matthew from joining the navy and also from applying for admission to West Point Military Academy. Despite this, the politician Sam Houston (1793–1863) obtained a warrant for Matthew to enter the navy as a midshipman. No naval academy existed at that time, and naval officers were trained at sea. In 1825 Maury was sent to sea in the 44-gun frigate *Brandywine*, which was carrying the marquis de Lafayette (1757–1834) home to France. Midshipmen were required to read a standard textbook, *New American Practical Navigator* by Nathaniel Bowditch. Maury read the book but found it did not answer all his questions about navigation, and he questioned its ability to teach navigation.

The *Brandywine* set sail again in 1826 for a voyage around Cape Horn. When they reached the South Pacific, Maury was transferred to the *Vincennes,* and in 1829, when the ship was ordered home by the long route, he began his journey around the world. He found that the *Vincennes* had an excellent library, and he spent much of his free time reading books on navigation, mathematics, trigonometry, and spherical geometry. During the long voyage Maury made notes on the winds and currents they encountered. In March 1831 Maury sailed again, this time as sailing master and navigator on the sloop of war *Falmouth.* They were bound for Cape Horn and, again, Maury recorded winds, tides, and currents, and each day he calculated the distance the ship had traveled. When they reached the Cape, the ship met strong westerly winds, but Maury calculated that farther south the winds would be from the east. They sailed south, found the easterlies, and reached Valparaíso, Chile, in 24 days.

After returning to the United States, Maury wrote *A New Theoretical and Practical Treatise on Navigation,* which was published in Boston in 1836. Nathaniel Bowditch, author of the textbook Maury found so unsatisfactory, praised it and recommended to the naval authorities that it be used instead of his own book to teach midshipmen.

Maury was then assigned to the brig *Consort.* The ship was undergoing repairs, so he took the opportunity to visit his parents. On the way back the stagecoach in which he was traveling overturned. Maury broke his right leg in two places and had to wait more than an hour for a doctor to arrive. The doctor set the fractures so badly that a second doctor had to break the leg and reset the bones. The result was that Maury would never again be able to command a ship. While recuperating he made several suggestions for naval reform, one of them being the establishment of a naval academy to train officers.

In 1841 Maury was placed in charge of the Depot of Charts and Instruments, a post he held until 1861. The Depot held the ships' logs that captains handed to the authorities at the end of every voyage, and Maury used these day-by-day records to develop an overall pattern of winds, tides, and currents. This was eventually assembled into a series of charts. In 1850 Maury charted the floor of the North Atlantic Ocean to facilitate the laying of the first transatlantic cable.

His study of winds and the pressure systems that give rise to them meant that Matthew Maury was a marine meteorologist as well as an authority on wind-driven ocean currents. He realized that meteorology called for international cooperation, and he urged the formation of an international weather service. He was unable to obtain government funding for this project, but in 1853 King Leopold of Belgium arranged a conference to discuss the idea. Several nations sent representatives, and Maury attended on behalf of the United States. The nations agreed to share meteorological data gathered to a uniform standard. In 1855 Maury published the first textbook on oceanography, *Physical Geography of the Sea,* and in the same year he was promoted from lieutenant to commander.

The Civil War broke out in 1861 and on April 20, three days after Virginia had seceded from the Union, Matthew Maury resigned his commission and joined the Confederate States Navy with the rank of commander. He was placed in charge of coastal, harbor, and river defenses. During the war he invented an electric torpedo and experimented with electric mines. He visited England, Ireland, and France, purchasing ships and supplies. After the war Maury went into voluntary exile. He lived for a time in Mexico, where he served as Imperial Commissioner for Immigration, building colonies for Confederate and other immigrants, and in 1866 he moved to England. He returned to the United States in 1868 to take up the post of professor of physics at the Virginia Military Institute, settling in Lexington. In the same year he helped to launch the American Association for the Advancement of Science.

Matthew Maury died at home in Lexington on February 1, 1873. He was buried temporarily in Lexington cemetery, but in 1874 his body was reburied in Hollywood Cemetery in Richmond, Virginia, between the graves of U.S. presidents James Monroe and John Tyler. In 1930 Matthew Maury was elected to the Hall of Fame for Great Americans.

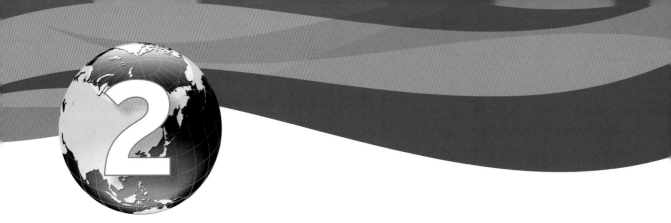

The Origin of the Oceans

Water finds its own level. When it flows across a surface with only the force of gravity acting upon it, water invariably flows downhill and accumulates in the lowest-lying areas. The rule that applies to water spilled on the floor and streams that flow into lakes also applies to the oceans. They fill basins, which are regions of the Earth's surface at a lower elevation than dry land. This much is obvious, but it raises two questions: Why is the Earth's surface so uneven, with high mountains and deep basins? Where did the water come from to fill the ocean basins?

Straightforward observation seemed to supply an adequate answer to the second question. As recently as the 16th century many scholars believed there were vast quantities of water below ground. Indeed, this seemed to be self-evident. Every village had at least one well, so everyone knew that if it were deep enough a hole dug into the ground would fill with water. There must be water below ground for this to happen. People also knew that water emerges spontaneously from the ground at springs and seeps—patches of wet ground where water rises to the surface. Miners often had to wade through water that flowed into mine galleries. In some places geysers shot steam and hot water high into the air. Water erupted from volcanoes. All of this water must have a source, so people imagined there was a vast, worldwide, subterranean reservoir.

René Descartes (1596–1650), the renowned French philosopher and mathematician, refined this idea and explained how the reservoir could be constantly replenished. He proposed that the Earth was constructed in five layers, as five concentric spheres. The Earth had once been molten, and the layers formed as the rocks cooled and solidified. One of the layers, according to Descartes, consisted of water. This was the ultimate source of all the Earth's water. He suggested that seawater moving through underground channels accumulated in large caverns below the tops of mountains. Water from the mountain caverns percolated through gaps in the rocks to emerge as springs that fed mountain streams, thus returning the water to the sea.

Scholars assumed that the water itself had always been present. The fact that volcanoes eject water vapor suggested that as the originally molten Earth cooled, water vapor condensed and rain filled the ocean basins. In fact, this cannot be the whole story, because physicists have calculated that even if the entire atmosphere consisted of water vapor it would be insufficient to fill the ocean basins to the depths of the present oceans. The modern view is that Earth formed through the accretion of gas, dust, and masses of rock from a disk of material. Gravitational pressure heated the accreting material, and the decay of radioactive elements also released heat. As the young Earth grew hotter, chemical reactions led to the formation of water that filled the low-lying basins.

This chapter deals not with the water of the oceans, however, but with the origin of the basins the oceans fill. For many years the prevailing view was that these resulted from the contraction of a cooling Earth. This theory became very elaborate, but in the end it was challenged and found wanting, and alternative explanations replaced it.

THE CONTRACTING EARTH

Today the rocks on land and the waters of the ocean are cool, but deep below the surface the Earth is hot. When miners first began digging deep tunnels, they noticed that the deeper they went the warmer the surrounding rocks were. People also knew that volcanic eruptions eject molten rock from beneath the surface. Today scientists know that while a little of the Earth's internal heat originated with gravitational compression during the planet's formation, most is generated by the decay of radioactive elements, but this is recent knowledge. Prior to the discovery of radioactivity by the French physicist Antoine-Henri Becquerel (1852–1908) in 1896, the only plausible explanation for the high temperature was that it was residual. Originally, long ago, the Earth had been molten, and it had been cooling throughout its history, from the exterior inward. The surface rocks had solidified, just as molten lava cools and solidifies, and formed a layer of solid rock that partly insulated the molten rock beneath, reducing the rate at which it cooled. But the cooling was—and is still—continuous.

Most substances contract when they cool. When coopers make barrels, they heat the metal hoops, causing them to expand, and fit the hot hoops over the wooden staves. As the hoops cool, they contract, forcing the staves tightly together and producing a watertight container. There was no reason to suppose that the Earth would behave differently. As it cooled, so it would contract, and its contraction would exert powerful horizontal forces, shrinking the solid crust.

Jean-Baptiste-Armand-Louis-Léonce-Élie de Beaumont (1798–1874), one of the most distinguished geologists of his generation, was the leading proponent of this view, using it to explain the origin of mountain ranges. He had studied 21 mountain chains in Europe and North America and discovered that many on either side of the Atlantic were geologically related, comprising similar suites of minerals and similar fossils. Élie de Beaumont argued that as the Earth cooled and contracted, its volume would decrease, and this would leave the outermost layer—the crust—unsupported. The crustal rocks would then crumple in a series of episodes that produced long chains of mountains along *great circles*. A great circle is an arc of a circle drawn on a sphere with its center at the center of the sphere. Not all geologists agreed with Élie de Beaumont, but one who did was Louis-Constant Prévost (1787–1856), the professor of geology at the Paris Athenaeum. Prévost maintained that as the Earth contracted, its crust would wrinkle like the skin of a stored apple as it dries.

Shrinking generates lateral forces, producing mountains and depressions the way a tablecloth crumples when pushed from two sides toward the center of a table. This was fundamentally different from the opposing view, which held that mountains form when material rises from below, thrusting through the crust the way *magma* rises to fill the chamber beneath a volcano. The theory that mountains form by the cooling and solidification of molten material rising from below the crust was known as Plutonism, after Pluto, the Roman god who dwelled in the underworld. The Plutonists were opposed by the Neptunists, after Neptune, the Roman god of the sea. The Neptunists believed that the entire surface of the Earth had once lain beneath water. That would explain how sedimentary rocks, which are known to form only from sediments deposited on the sea-bed, came to be found on mountains and in the interior of continents, far from the nearest coast. It was not until the development of the theory of plate tectonics (see "Plates, Ridges, and Trenches" on pages 79–84) that this dispute was finally resolved.

JAMES DWIGHT DANA AND THE ORIGIN OF OCEAN BASINS

James Dwight Dana (1818–95), the most distinguished American geologist of his generation, agreed with Prévost that the Earth was contracting and, as it contracted, that the crust crumpled. Geologists calculated that the Earth had been cooling for approximately 100 million years, and that during that time its circumference had shrunk by several hundred miles. Dana used this theory to explain how the geologically complex Appalachian Mountains formed.

Dana's development of the contracting Earth theory also explained the formation of ocean basins. Dana began with the metaphor of the crumpling tablecloth, but he believed that this description of the process was incomplete. The surface of the Earth would not cool evenly, which implied that the lateral forces resulting from the shrinking and crumpling the crustal rocks would not be the same everywhere. At the same time, the material below the crust would also cool unevenly. This would produce large depressions, where the underlying material cooled and contracted vertically relatively rapidly and blocks of crustal rock sank with it. Dana believed that the section of crust lying between North and South America in the west and Europe and Africa in the east sank in this way, forming an extensive depression that filled with water, producing the North and South Atlantic Oceans. Dana held that this stage of contraction occurred early in the history of the Earth and, consequently, that the continents and oceans are very ancient.

Contraction did not end, however, and the lateral compression of the crust continued. The continents were exposed to erosion by wind, rain, and expansion and contraction due to the summer heat and winter cold. Rivers transported the eroded rock particles, depositing them in the sea where they settled to the seabed, forming sediments that hardened into sedimentary rock. Crustal compression then crumpled the sedimentary rocks, raising mountain ranges consisting of sedimentary rocks along continental coasts. That is how Dana explained the formation of the Appalachians. Scientists now know that lateral compression does occur and that it can raise mountain chains, so Dana was partly correct. Where he was wrong was in his view that once they had formed, the continents and ocean basins became permanent features of the Earth's surface. They are not: Ocean basins open, widen, and then close again, and continents move, sometimes joining to form *supercontinents* that later break into sections that move apart.

James Dwight Dana was born in Utica, New York, on February 12, 1813. His father, James Dana, owned a successful hardware store, and the young James was the eldest of his 10 children. James's mother, Harriet Dwight Dana, was deeply religious and a strong influence in his life. When he was 14, James enrolled at Utica High School, where Fay Edgerton, one of the teachers, encouraged his interest in science. In 1830, at the age of 17, he entered Yale College, where he studied under the chemist Benjamin Silliman (1779–1864). Dana graduated from Yale in 1833. He had exhibited a talent for mathematics and a keen interest in botany and mineralogy, but it was not easy in those days to earn a living as a scientist, and Dana became a teacher, obtaining a post as a naval schoolmaster. There was no naval college to train officers, and midshipmen learned their seamanship at sea. Dana was assigned to the USS *Delaware*, sailing to the Mediterranean, and in 1834 he returned on the frigate *United States.* During the voyage he made a number of scientific notes and observations, and a description of an eruption of Vesuvius that he sent to Professor Silliman was his first scientific paper, published in the *American Journal of Science*, which Silliman edited.

After his return, Dana spent part of his time at the family home in Utica and part in New Haven, Connecticut. He had no immediate prospect of employment so continued studying chemistry and mineralogy, until in 1836 he obtained a position as assistant to Silliman. The work occupied him for only three hours each day, and during the two years he remained in the post, Dana developed a system for classifying minerals by their chemistry and crystallography. He published this in 1837 with the title *System of Mineralogy and Crystallography.*

He was not yet 24 years old, and the book established his scientific reputation.

It was then that Dana became an explorer. The government was being urged to send an exploratory expedition to the South Pacific, where several other nations were showing interest. An expedition was authorized in 1836. Dana was invited to take part, but at first he refused. The botanist Asa Gray (1810–88) persuaded him to reconsider, and in January 1837 Dana was commissioned as its official mineralogist. The expedition sailed from Norfolk, Virginia, on August 18, 1838, under the overall command of Lieutenant (later Rear Admiral) Charles Wilkes (1798–1877). It comprised two sloops of war, the *Vincennes*—the ship on which Matthew Maury had sailed as a midshipman in 1829—and *Peacock,* and four smaller supply ships and tenders. The ships sailed to Madeira, St. Iago in the Cape Verde Islands, Rio de Janeiro, and from Rio to Cape Horn. After rounding the Cape and weathering its storms, the expedition continued to Valparaíso, Chile, and Callao, Peru, and then into the South Pacific, calling at Tahiti, Samoa, and Sydney, New South Wales, Australia. While Dana was conducting his researches in Australia and New Zealand, Wilkes headed south to Antarctica. In September 1840 the expedition sailed to Tonga and Hawaii. After many adventures Dana finally reached San Francisco in October 1841. The ships then sailed from California to Honolulu, Singapore, the Philippines, around the Cape of Good Hope, and to Saint Helena, and the *Vincennes,* with Dana on board, reached New York Bay on June 10, 1842. Dana then settled in Washington, D.C., where he began work on his reports of the voyage.

Writing the reports occupied most of the next 13 years, but Dana was not happy in Washington. In 1844 he moved back to New Haven, where on June 5 he married his fiancée, Henrietta Frances Silliman, the daughter of his professor. In 1846 Dana became chief editor of the *American Journal of Science,* and in 1850 he succeeded Benjamin Silliman at Yale, becoming Silliman Professor of Natural History and Geology, a post he held until 1892.

James Dana was a very eminent scientist, and most other scientists supported his theory of the way that mountains and ocean basins form. He believed the process also affected living organisms. As compressive forces generated by contraction of the crust added more mountains to the edges of continents, the continents became larger and their climates harsher. That drove organisms to become more complex. Dana saw this as evidence that what he called the "Power Above Nature" had prepared the Earth for humans, the most complex of all organisms and, in his view, the purpose and endpoint of history.

James Dana took on a tremendous workload, teaching, writing, revising earlier works for new editions, and editing and contributing to the *American Journal of Science.* It was too much for any man, and in 1859 he collapsed from exhaustion. He spent 10 months recuperating in Europe but never fully recovered. There were repeated relapses, and he had to strictly curtail his social life. He continued working, however, until he died peacefully on April 14, 1895.

OSMOND FISHER AND THE ORIGIN OF THE PACIFIC BASIN

Most geologists agreed with James Dana that the positions and general shapes of the continents and oceans were permanent—but not all. Others accepted Dana's explanation for the formation of the Atlantic Basin but found evidence suggesting that mountains formed in ways other than the mechanism Dana had proposed for the Appalachians. One of Dana's most prominent opponents was the Austrian geologist Eduard Suess (1831–1914), who occupied a position of eminence in Europe comparable to that occupied by Dana in North America.

Suess spent many years studying the Alps, and he concluded that these mountains had formed fairly recently as a result of lateral compression caused by land to the south moving slowly northward. This was compatible with Dana's theory up to a point, but the Alps had formed not by folding but by fracturing and overthrusting—the process in which large slabs of rock are pushed above and over those ahead of them. Suess also proposed that the erosion of continental rocks produced material that accumulated on the seafloor as sediments and that the sediments eventually filled the ocean basins. The thickening layer of sediment raised the sea level until the sea invaded low-lying areas of land, forming inland seas. From time to time a section of continental crust would collapse, forming a new basin that would fill with water from the inland seas.

The process Suess described differed greatly from the one Dana had proposed, and it led Suess to a dramatic conclusion. He suggested that there had been a time when South America, Africa, India, and Australia had all been joined together as a single supercontinent, which he called Gondwanaland. He took the name from Gondwana, a geographical region of central India where there were fossils similar to those found in the constituent continents of Gondwanaland. After a long time the interior of Gondwanaland collapsed. Sea flooded in, dividing the supercontinent into the present-day continents. Eduard Suess also proposed the existence of the Tethys Ocean, which has since disappeared (see "Tethys, Panthalassa, and the Drifting Continents" on pages 72–78).

Suess challenged Dana's idea that the continents and oceans are permanent, and both theories attracted supporters. Regardless of their views on the permanence of continents and oceans, however, most agreed that the Earth had once been molten, that it was cooling and contracting, and that mountains formed because of compressive forces generated by contraction.

Physicists, however, calculated that the compressive forces produced by contraction would be too weak to crumple solid rocks into high mountains. Modern geophysicists have additional objections to the theory. If the entire planet contracted, it would conserve

CONSERVATION OF ANGULAR MOMENTUM

A body rotating about its own axis possesses three properties: mass (M); the radius (r) of the circle it describes; and its angular velocity (V). Angular velocity is the speed with which the body moves along a circular path; it is usually measured in radians per second (rad/s). Angular momentum is the tendency of a rotating body to continue moving along its circular path. Its magnitude is the product of the body's mass, angular velocity, and radius of rotation: MVr.

Angular momentum is conserved. That is to say, it can be transferred to another body, but provided no outside force (such as friction) acts to increase or decrease it, angular momentum will remain constant. Consequently, if one component of angular momentum changes, one or both of the other components will automatically change to compensate. Ordinarily, it is impossible for the mass (M) of a rotating body to change, but it is possible for its angular velocity (V) and rotational radius (r) to change, and when one of these changes the other must also change.

Suppose a body wherein $M = 1$; $V_1 = 5$; and $r_1 = 20$. Then $L = MV_1r_1 = 1 \times 5 \times 20 = 100$. Suppose the radius is halved, so $r_2 = 10$. Then $MV_2r_2 = 100$ (because L remains constant). Therefore $V_2 = 100/Mr_2 = 10$. If the radius is decreased by half, the angular velocity doubles to compensate.

its angular momentum as it did so, and this would have made it spin faster on its axis (see the sidebar). There is no evidence that this happened. If mountains formed by crumpling, they should be distributed fairly evenly over the Earth's surface, but they are not. Mountains occur in particular ranges and chains. And compression is not the only force at work. Geologists have identified large-scale features that were produced by tension—pulling apart—rather than compression—pushing together. Rifts, such as the East African Rift System that includes the Great Rift Valley form when the crust fractures and two sections move apart with a deep depression, called a *graben,* between them. More recently still geophysicists have discovered that radioactive decay is the principal source of the Earth's internal heat and, therefore, that the Earth is not cooling from an original molten state.

Osmond Fisher (1817–1914) was one geophysicist who disagreed strongly with the idea that contraction could produce enough force to raise mountains. In the late 19th century geophysicists were debating the thickness of the Earth's crust. It was clear that if the temperature continued to rise with increasing depth at the average rate measured in mines, then rock would be molten below a depth of about 20 miles (32 km). But if the Earth's interior were liquid, it would be subject to tidal forces and the surface would rise and fall with the tides. One consequence of that would be the absence of ocean tides at coasts, because the land would be rising and falling at the same time as the water. Fisher proposed that the interior of the Earth is molten, but that it is able to expand and contract and is in constant motion. He estimated the average density of solid rock and the molten interior and calculated from this that at sea level the crust was 18 miles (29 km) thick and that the motion of the liquid interior prevented it from growing any thicker. In explaining how tides could affect the oceans but not the continents, Fisher had come close to discovering a mechanism for the movement of continents.

This was not Fisher's only contribution to the scientific understanding of the Earth. He had an explanation for the formation of the Pacific Basin. This arose from another topic of discussion in the 19th century: the origin of the Moon. Sir George Howard Darwin (1845–1912) suggested in 1878 that Earth had formed by condensation from a cloud of gas and dust, and for a time during its formation, the planet had been liquid. It was also spinning rapidly, and Darwin had calculated that the Sun's gravitational attraction had combined with inertia—the tendency of a moving body to continue moving in a straight line—to produce a bulge around the equator and had overcome Earth's own gravitational force at that bulge. A stream of material had flown away from the Earth and condensed to form the Moon. Tidal forces between the Earth and Moon then placed the Moon in its orbit and synchronized its rotation so it always presents the same side to Earth.

Fisher's contribution was to propose that the material leaving Earth, eventually to form the Moon, left behind a deep depression, like a vast hole in the ground from which material had been excavated. That depression later filled with water to become the Pacific Ocean. In 1882 he described this process in "On the Physical Cause of the Ocean Basins," a paper published in *Nature.* He followed this in 1892 with a further *Nature* paper, "Hypothesis of a Liquid Condition of the Earth's Interior Considered in Connexion with Professor Darwin's Theory of the Genesis of the Moon." Unfortunately, his ingenious idea was unable to account for the fact that the Moon's orbit is at an angle to the plane of the Earth's equator. If the Moon originated in material taken from an equatorial bulge, it should orbit in the equatorial plane. In the 1930s the English astronomer and geophysicist Sir Harold Jeffreys (1891–1989) finally demolished this theory of lunar origin by showing that friction inside the liquid Earth would dampen the force producing an equatorial bulge and prevent the ejection of material in the way Darwin had proposed.

Osmond Fisher was born at Osmington, Dorset, a county in the south of England, on November 17, 1817. He was a Church of England (Episcopalian) clergyman for all his adult life, but in common with many clergy he devoted his free time to scientific pursuits. He studied the geomorphology of Norfolk, in eastern England, and the sedimentary rocks and fossils of Dorset. Fisher died at Huntingdon on July 12, 1914.

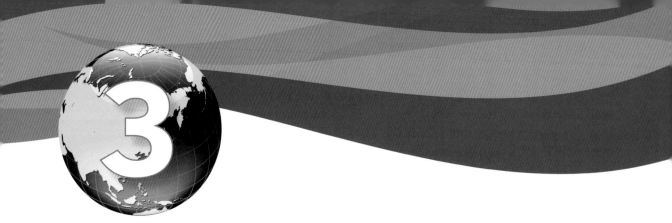

Studying the Ocean Floor

By the second half of the 19th century cartographers had mapped the oceans and hydrographers had surveyed and charted most of the world's coasts and islands. The oceans were known, their currents marked on nautical charts and in school atlases. A century later, orbiting satellites were gathering data on the height of sea level, the temperature of the water, and even on the amount of biological activity in the oceans. No remote atoll, no headland or cape, no river estuary or iceberg could hide from this constant surveillance.

It might seem that the oceans had yielded all of their secrets, but this was very far from the truth. The ocean surface had been explored and charted, but little was known about what might lie beneath the surface, and the ocean floor was as remote and inaccessible as a distant planet. Oceans cover 70.8 percent of the Earth's surface, their average depth is 2.32 miles (3.73 km), and it was not until late in the 19th century that scientists were able even to begin the exploration of the ocean floor and its living inhabitants.

This chapter tells of the exploration of the ocean floor. The story begins in 1872 with the start of the world's first oceanographic expedition on HMS *Challenger*, a converted Royal Navy warship, whose scientists were the first to sample sediments from the ocean floor. Although divers were able to move and work on the seafloor, they were confined to shallow water. Scientists seeking to map the floor of the deep ocean had to devise ways of doing so working from ships on the surface. The chapter describes how they achieved this.

As the technologies for exploring the ocean floor advanced, scientists discovered submarine mountains, volcanoes, and trenches—undersea canyons with sides that plunged to immense depths. The picture of the ocean floor that slowly emerged led scientists to the realization that ocean basins are not permanent. Over millions of years they open, widen, then grow narrower and finally disappear. This led to the discovery of plate tectonics, the theory that now underpins all of the geological sciences.

HMS *CHALLENGER*

The summit of Mount Everest, the world's highest mountain, is 29,029 feet (8,848 m) above sea level. The bottom of the Mariana Trench, on the ocean floor, is 36,091 feet (11,000 m) below sea level. Mount McKinley, the tallest peak in North America, rises to 20,320 feet (6,194 m). The abyssal plain lies up to 20,000 feet (6,000 m) below the surface and covers approximately 40 percent of the ocean floor. Only fit and experienced climbers are able to reach the peaks of the world's highest mountains, but the abyssal plains and oceanic trenches are much more inaccessible still. The diagram shows a cross-section of the ocean. It greatly exaggerates distances and gradients, but it shows just how far below the surface the deep ocean floor lies.

Exploration of the ocean floor began with a voyage of discovery that would become the most famous oceanographic expedition ever launched. On December 21, 1872, under the command of Captain (later Admiral Sir) George Strong Nares (1831–1915), HMS *Challenger* sailed from Portsmouth, England, with a crew of 20 officers, 205 sailors, and six scientists. Led by Charles Wyville Thomson (1830–82; see "Charles Wyville Thomson" on pages 48–52), the scientific team consisted of Scottish chemist John Young Buchanan (1844–1925), English zoologist Henry Nottidge Moseley (1844–91), Scottish-Canadian zoologist John Murray (1841–1914), German zoologist Rudolf von Willemoes-Suhn (1847–75), and Swiss artist Jean-Jacques (later John James) Wild (1828–1900). The *Challenger* arrived back in England on May 21, 1876, having sailed 68,890 nautical miles (79,224 miles; 127,471 km). She had crisscrossed every ocean except the Indian Ocean and had visited North and South America, South Africa, Australia, New Zealand, Hong Kong, Japan, and many islands in the Atlantic and Pacific Oceans. Wherever they went ashore, the

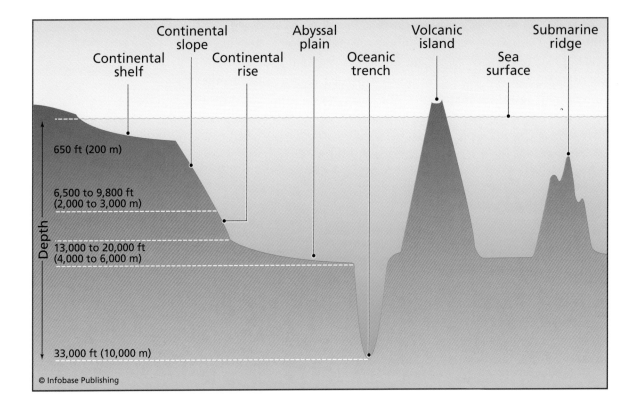

Continental shelf

Continental slope

Continental rise

Abyssal plain

Oceanic trench

Volcanic island

Sea surface

Submarine ridge

Depth

650 ft (200 m)

6,500 to 9,800 ft (2,000 to 3,000 m)

13,000 to 20,000 ft (4,000 to 6,000 m)

33,000 ft (10,000 m)

© Infobase Publishing

This cross-section of the ocean floor greatly exaggerates vertical distances and gradients and compresses horizontal distances. The abyssal plain covers most of the ocean floor. It lies 13,000–20,000 feet (4,000–6,000 m) below the sea surface.

expedition scientists collected botanical and zoological specimens and studied many of the peoples they encountered. Wild, the official artist, drew many of the specimens and the events where members of the expedition met important persons—including the king of Portugal and emperor of Japan—but the team also took many photographs. Photography was still new, and this was probably the first expedition to make routine use of it. In all, the *Challenger* spent 713 days at sea, and that is where her most important work was done.

Every two or three days the ship would stop at one of the 362 official stations where the scientists would measure the depth of water and take samples of the seabed sediment. They measured the temperature at the surface, at the seabed, and at several intermediate depths, and they collected water samples for chemical analysis. Their measurements of temperature disproved what was then the widely held view that water near the ocean floor was at a constant 39°F (4°C), based on the false assumption that seawater is at its densest at the same temperature as freshwater. At average salinity of 35‰ (parts per thousand), seawater is densest at 32°F (0°C) and the average water temperature at the abyssal plains is 33–36°F (1–2°C). Then the crew would collect biological samples using a trawl or dredge dragged across the seabed and plankton nets drawn through the water to a depth of 4,921 feet (1,500 m). Thomson reported that when the voyage ended they had on board: "563 cases, containing 2,270 large glass jars with specimens in spirit of wine, 1,749 smaller stoppered bottles, 1,860 glass tubes, and 176 tin cases, all with specimens in spirit; 180 tin cases with dried specimens; and 22 casks with specimens in brine." These were in addition to more than 5,000 bottles and jars of all sizes they had sent from various parts of the world to be stored at Edinburgh—of which only about four were broken.

The *Challenger* soundings of depth concentrated on the deepest parts of the oceans. It was the first expedition to obtain sediment samples—25 of them—from below 14,765 feet (4,500 m), including one from 18,701 feet (5,700 m). The deepest sounding reached 26,850 feet (8,189 m) in the southwestern Pacific at a site that is now known as the Challenger Deep.

The British government paid the cost of the *Challenger* expedition. This amounted to about £200,000, equivalent to more than £10

Scientists at work during the cruise (1872–76) of the British survey ship HMS *Challenger* (Granger Collection)

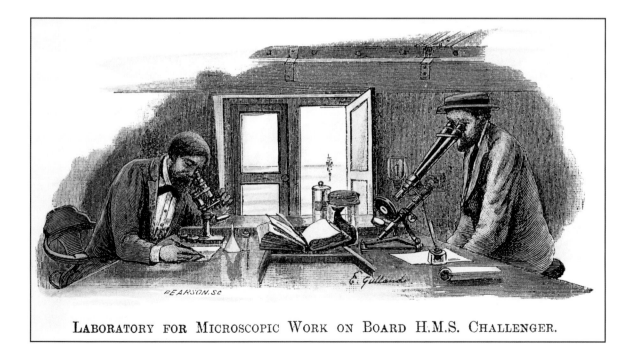

LABORATORY FOR MICROSCOPIC WORK ON BOARD H.M.S. CHALLENGER.

million ($20 million) today. The ship, HMS *Challenger*, was a three-masted, square-rigged, naval *corvette*, launched in 1858. She was 200 feet (61 m) long and displaced 2,306 tons (2,343 tonnes). As well as sails, the ship had a steam engine. Originally the *Challenger* was fitted with 17 guns, but 15 of these were removed for the oceanographic expedition. Her *spars* were also shortened so they occupied less space when stored. To prepare for the expedition the *Challenger* was fitted with winches and a special deck to support them, laboratories, civilian accommodation, and 249 miles (400 km) of rope.

After the expedition, the *Challenger* spent some time working for the coast guard and then as a training ship based at Harwich, on the east coast of England. She remained in reserve until 1883 and then became a *receiving hulk.* She was finally broken up in 1921. Nothing remains of her except her figurehead, which is now displayed in Southampton.

CHARLES WYVILLE THOMSON, SCIENTIFIC LEADER OF THE *CHALLENGER* EXPEDITION

The *Challenger* expedition was the brainchild of two British scientists, Charles Wyville Thomson (1830–82) and William Benjamin Carpenter (1813–85). Carpenter was a senior member of the Royal Society and wielded considerable influence in official circles, which helped them promote their project. Carpenter did not sail on the *Challenger*, however, feeling that at 59 he was too old for such a venture. Their aim was to discover whether living organisms inhabited the deep ocean floor. The prevailing view among biologists, with which Thomson by that time disagreed, was that marine life was confined to the upper layers of water and that below 200 fathoms (1,200 feet; 366 m) the sea was lifeless. The idea that nothing can live below 200 fathoms is known as the *azoic* hypothesis, and it was first proposed in 1843 by the British naturalist Edward Forbes (1815–54).

Edward Forbes was born on February 2, 1815, at Douglas, the principal city in the Isle of Man, an island in the northern Irish Sea between Northern Ireland and England, that is a self-governing dependency of the British crown; it is not part of the United Kingdom and is not a member of the European Union. As a child Forbes read avidly, drew caricatures, and collected natural history specimens of all kinds. Until he was 13 his health was too poor for him to attend

Forbes fell ill. He died at Wardie, near Edinburgh, on November 18, 1854, aged only 39.

Whether Forbes believed in the literal truth of his azoic hypothesis is uncertain, but within a few years of his death contradictory evidence had started to emerge. The eminent Norwegian biologist Michael Sars (1805–69) was certain Forbes was mistaken. He conducted dredging operations in Norwegian fjords, reaching depths approaching 3,300 feet (1,000 m), and in 1868 he listed 427 species of animals found at depths of about 450 fathoms (2,700 feet; 823 m). In 1864, in a haul dredged from about 1,800 feet (550 m) in Lofoten, he and his son George Ossian Sars found a specimen of *Rhizocrinus lofotensis*, a "sea lily." This was the first living example of a stalked crinoid, previously known only from fossils. Crinoids (class Crinoidea) are primitive echinoderms—the phylum that includes starfish and sea urchins. The illustration, based on a drawing made during the *Challenger* expedition, shows what *Rhizocrinus* looks like.

Charles Wyville Thomson knew of Sars's discoveries and was already interested in crinoids. He and Carpenter persuaded the authorities to allow them to use the warships *Lightning* and *Porcupine* for several deep-sea dredging expeditions. *Lightning*, a temperamental paddle steamer in poor condition, spent the summer of 1868 dredging at depths of 300–3,900 feet (90–1,189 m) and found abundant life. The following year the Admiralty allowed Thomson and Carpenter to conduct four cruises in *Porcupine*. They explored beyond the depths reached by *Lightning*, one dredge reaching a depth of 14,072 feet (4,289 m). Again they found abundant life. This finally laid the azoic hypothesis to rest. It was the success of the *Lightning* and *Porcupine* cruises that made it possible for the two scientists to win support for the *Challenger* expedition.

Thomson was christened Wyville Thomas Charles Thomson. He changed his name to Charles Wyville Thomson in 1876, when he was knighted. He was born on March 5, 1830, at Bonsyde, Linlithgowshire (now West Lothian). His father was a surgeon employed by the British East India Company. Charles was educated at Merchiston Castle School, a private school in Edinburgh, and at the University of Edinburgh. In 1850 he was appointed to a lectureship in botany at the University of Aberdeen, and in 1851 he became professor of botany there. In 1853 he moved to Ireland, as professor of natural history at Queen's College, Cork (now University College Cork and

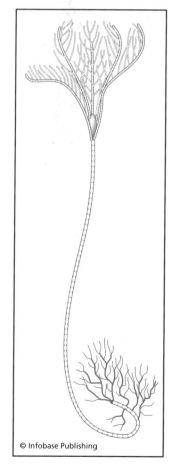

Crinoids, such as *Rhizocrinus lofotensis* shown here, look like plants, but they are animals, belonging to the same phylum as sea urchins and starfish. They live anchored to the seafloor, using their tentacles to trap particles of food.

© Infobase Publishing

part of the National University of Ireland). Thomson was there only one year before moving to the Queen's University of Belfast as professor of mineralogy and geology. In 1860 he transferred to professor of natural history at the same university. He became professor of botany at the Royal College of Science for Ireland, in Dublin, in 1868, and in 1870 he was appointed Regius Professor of Natural History at the University of Edinburgh—the chair formerly held by Edward Forbes.

On his return from the *Challenger* expedition, Thomson received many academic honors. His health had never been robust, however, and in the 1870s it deteriorated. Thomson gave no lectures after 1879, and he died at his Bonsyde home on March 10, 1882.

CHARLES BONNYCASTLE AND THE DREAM OF CHARTING THE OCEAN FLOOR

Nowadays when oceanographers wish to study the ocean floor, they often use equipment that emits a sound wave vertically downward or to the side. When the sound wave strikes an object or surface, it is reflected, and the equipment detects the reflection and measures the time that elapses between the transmission and the reception of the reflection—the echo. The time lapse, measuring the time taken for the sound wave to reach the reflector and return from it, indicates the distance to the reflector, and small variations in the time lapse across the width of the signal reveal the shape of the reflecting surface. This is called *echo sounding*—sound waves directed to the side are used in *side-scan sonar.* The speed of sound varies according to the density of the medium through which the sound waves travel. In seawater with the average salinity of 34.85‰ and at 32°F (0°C), sound travels at 3,232 mph (1,445 m/s), increasing by about five mph for every 1°F rise in temperature (4 m/s per 1°C), by about 3.3 mph (1.5 m/s) for every 1‰ increase in salinity, and by about 12 mph for every 300 feet increase in depth (18 m/s per 1,000 m). Echo sounding is highly accurate.

In November 1826 a Swiss physicist, Jean-Daniel Colladon (1802–93), and his assistant Jacques-Charles-François Sturm (1803–55) measured the speed of sound in water by an experiment they conducted on Lake Geneva. Sturm sat in a boat and lowered a church bell weighing 140 pounds (64 kg) a few feet into the water. He struck the bell with a hammer, at the same moment firing a flare into the sky. Colladon sat in a second boat about 8.6 miles (13.887 km) away, listening for the sound of the bell through an ear trumpet 17 feet (5 m) long. The ear trumpet was a tin tube. The end submerged below the surface was sealed, and the other end opened into a conical shape, covering Colladon's ear. Colladon heard the sound of the bell, noted the time that elapsed between seeing the flare and hearing the sound, and from this Colladon and Sturm calculated the speed of sound in water at a temperature of 46°F (8°C). They found it to be 3,210 mph (1,435 m/s). The two friends received an award from the Académie des Sciences in Paris for their very accurate measurement.

Some years later Colladon repeated the experiment on Lake Geneva, this time with a bell weighing 1,100 pounds (500 kg). This time he heard the sound clearly over a distance of almost 22 miles (35 km). Interestingly, one of Colladon's companions who was sitting in a boat about two miles (3 km) from the bell heard a second sound, which was apparently an echo from the lake bank behind the bell. They reported this but did not pursue it.

Across the Atlantic, at least one scientist was following Colladon's work closely. Since the University of Virginia first opened in 1825, Charles Bonnycastle (also spelled Bonneycastle; 1796–1840) had been its professor, first of natural history and of mathematics from 1827, when the first professor, Thomas Hewitt Key (1799–1875), returned to London. In 1838 Bonnycastle attempted to measure the depth of the ocean in the Gulf of Mexico by using reflected sound.

Bonnycastle sat in a boat 450 feet (137 m) from the *brig* USS *Washington,* while sailors on board the *Washington* detonated a small iron bomb called a *petard* below the surface. Bonnycastle listened with a trumpet similar to the one Colladon used but with the conical end immersed in the sea and pointing downward and the narrow end to his ear. He heard the explosion clearly, and one-third of a second later he heard a second, fainter sound, which he took to be an echo from the seafloor. Using the value for the speed of sound Colladon had measured, Bonnycastle calculated that the sea was 261 fathoms (1,569 feet; 478 m) deep. The *Washington* crew then measured the depth of water with a lead and found it to be approximately 540 fathoms (3,240 feet; 988 m). No one knows how such a large error occurred, but it is possible that the second sound Bonnycastle heard was an aftershock from the explosion, and he did not hear an echo from the seabed at all. The experiment failed, and Bonnycastle never tried to repeat it. His attempt at echo sounding was ahead of it time.

Charles Bonnycastle was born on November 22, 1796, in Woolwich, London, where his father, John Bonnycastle, was professor of mathematics at the Royal Military Academy. John was also the author of a textbook, *Introduction to Algebra,* and Charles helped his father prepare its 13th edition, published in 1824, adding this to the credentials he presented to the agent for the new state of Virginia. He published his own textbook, *Inductive Geometry,* in 1834. Charles Bonnycastle died on October 31, 1840.

REGINALD FESSENDEN AND HIS ECHO SOUNDER

The traditional way to measure the depth of the sea was to throw a weighted rope over the side of the ship and continue paying out rope until the rope went slack, indicating that the weight was resting on the bottom. This was called "taking a sounding." Knots tied in the sounding rope, or pieces of leather or other material fastened to the rope, at intervals of one fathom (6 feet; 1.8 m) meant the sailor taking the sounding had only to count the number of knots that passed through his hand. It was a simple system and perfectly adequate when all that the ship's master required was to know the depth of water below the ship to avoid the risk of grounding.

HMS *Challenger* used this method, but not only to ensure the vessel's safety in shallow waters. The *Challenger* crew took soundings to measure the depth of the open ocean, beyond the edge of the *continental shelf.* For this purpose they used the Baillie sounding sampler, shown in the illustration. The heavy weights at the end of the rope carried the rope to the ocean floor. When it struck the floor, the hollow tube at the center partly filled with sediment, and at the same time the wires attached to the top of the tube relaxed, dropping the weights onto the floor. Sailors then hauled the tube back to the ship, with its sample of sediment from a known depth.

At least, the sediment came from the ocean floor. The depth was less certain, because this technique for measuring depth is not very accurate in deep water. There are two reasons for this. The first is that the rope itself is heavy, especially when it is soaked in water, and its own weight will make it continue to sink long after the weighted end has come to rest on the seafloor. This will give an overestimate of the water depth. The second source of inaccuracy is that the ship continues to move while the weighted rope is sinking. Consequently, the rope does not descend vertically, but at an angle—perhaps quite

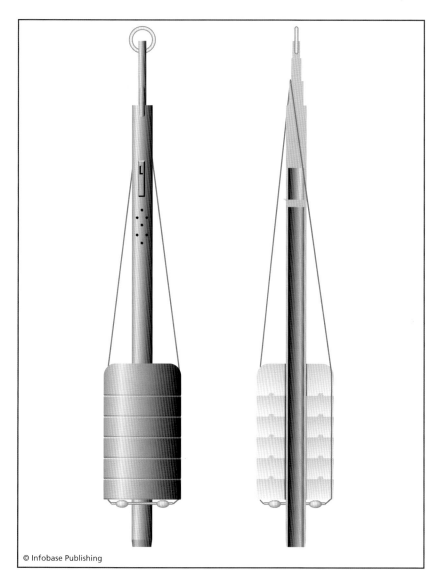

© Infobase Publishing

The Baillie sounding sampler was used on the *Challenger* expedition. The weights pulled the sounding rope to the sea bottom. When the device struck the bottom, the hollow tube filled with sediment, and the wires holding the weights relaxed, dropping the weights. The tube with its sample of sediment from a measured depth was then pulled up to the ship.

a steep angle—to the vertical. This also leads to an overestimate of depth.

Echo sounding is much more reliable, as Charles Bonnycastle believed but failed to prove. It was not until 1913, 75 years after Bonnycastle's attempt, that the German physicist Alexander Behm (1880–1952) invented an echo sounder that worked. He did not set out to measure depth, however, but to detect icebergs. On April 15, 1912, a collision with an iceberg drifting into the Atlantic from an Icelandic glacier sank the ocean liner *Titanic*. Behm hoped to prevent such a disaster from ever happening again. His echo sounder emitted a pulse of sound and used the time interval between emission and the arrival of the echo to calculate distance. In 1920 Behm established the Behm Echo Sounding Company in Kiel, Germany, to manufacture and market his invention. What his experiments revealed, however, was that icebergs did not reflect sound waves very well. On the other hand, the seafloor reflected them clearly, which meant echo sounding could be used to map the ocean floor.

There was another way to detect icebergs. Following the *Titanic* disaster, the Canadian physicist and inventor Reginald Aubrey Fessenden (1866–1932) announced that he had bounced radio signals off icebergs and had been able to measure the distance to them. This was a precursor of radar (an acronym derived from *ra*dio *d*etection *a*nd *r*anging). In 1915 Fessenden invented his own device for measuring sea depth by echo sounding. He called it the *fathometer*, from the words "fathom" and "meter," and patented it in 1917. During World War I Allied warships found Fessenden's fathometer very useful in detecting enemy submarines. *Scientific American* awarded him its gold medal for the invention in 1929.

Fathometers are still used, principally as *fish finders* that display on a screen echoes from shoals of fish and also for measuring sea depth. They can be used in a boat that is moving at cruising speed, but they provide only limited information about the topography of the seafloor. They work by converting an electronic signal into a pulse of sound that is transmitted vertically downward beneath the vessel and using a *hydrophone*—an·underwater microphone—to detect the echo. Most modern fathometers transmit a narrow bandwidth 200 kHz sound pulse.

Reginald Aubrey Fessenden was born on October 6, 1866, at Knowlton, Québec, the son of Elisha Joseph Fessenden and

Clementina Fessenden (born Trenholme). When Reginald was five years old, the family moved to Fergus, Ontario, and later to Chippawa, Ontario, close to Niagara Falls. He was educated at Trinity College School in Port Hope, and when he was 14, he was awarded a mathematics mastership to Bishop's College School in Lennoxville, Québec, which prepared boys for entry to Bishop's University, situated on the same campus. For a time Fessenden was studying at the university while also teaching mathematics to younger boys at the school. The mastership provided him with free tuition for one year if he passed the examinations, as well as a small income. Fessenden passed the examinations, but the curriculum centered on the classics, and he was more interested in science and especially radio, so at the age of 18 he left college to take up a teaching post at the Whitney Institute in Bermuda.

Fessenden was fascinated by the achievements of Thomas Alva Edison (1847–1931), and in 1886 he sailed from Bermuda to New York and obtained a job at Edison's laboratory in Menlo Park, New Jersey. Fessenden was especially interested in audio reception, and his work for Edison led to major improvements in radio receivers. He left the laboratory in 1890 when financial difficulties forced Edison to lay off most of the staff and spent a year working for several companies. He visited England at this time and was much impressed by the steam turbine he saw in Newcastle-upon-Tyne. He predicted, accurately, that electricity-generating steam turbines would one day propel ships of all sizes. In 1892 he was appointed professor of electrical engineering at Purdue University, and in 1893 he was recruited personally by George Westinghouse (1846–1914), head of the Westinghouse Corporation, to head the new electrical engineering department at the Western University of Pennsylvania (now the University of Pittsburgh).

Reginald Fessenden left the university in 1900 to work for the U.S. Weather Bureau, where he aimed to convert coastal weather stations from using telegraph lines to radio for transmitting their data. He invented several devices to improve audio reception, and on December 21, 1900, he successfully transmitted a voice message over a distance of 1.6 miles (1 km); this was probably the world's first voice broadcast. Fessenden left the Weather Bureau in 1902, but he had patented all of his inventions, and he retained the rights to them. In January 1906 he successfully exchanged radio messages in Morse code between Brant

Rock, in Marshfield, Massachusetts, and Machrihanish on the Kintyre Peninsula in the west of Scotland—at that time Guglielmo Marconi (1874–1937) had transmitted only one-way messages. On December 24, 1906, Fessenden broadcast a short program from Brant Rock to radio operators on ships near the Atlantic coast. Preceding it with a Morse signal indicating that an important message was to follow, he announced it himself, then played an Edison wax-cylinder recording of Handel's *Largo,* and ended the broadcast by wishing his listeners a happy Christmas and asking them to write to him with their impressions. The letters he received proved the broadcast had been a success. He transmitted a second program on New Year's Eve. These broadcasts were the first radio programs in the modern sense.

In 1928 Fessenden returned to Bermuda, where he had bought a small estate called Wistowe. He had become engaged to Helen Trott while he still lived in Bermuda. They were married in September 1890 and had one son, Reginald Kennelly Fessenden. Reginald Fessenden died at his home in Bermuda on July 22, 1932. Fessenden held more than 500 patents. In addition to his radio work and invention of the fathometer, he invented carbon tetrachloride (now banned because it is carcinogenic, but the original dry-cleaning fluid), the voice scrambler, the radio compass, the tracer bullet, the beeper-pager, and the automatic garage-door opener.

MID-ATLANTIC RIDGE

Part of the *Challenger*'s mission in 1872 was to explore the possible route for a future transatlantic telegraph cable. Several such cables already existed. After five failed attempts to lay them between 1857 and 1865, the first cable was laid successfully in 1866 by the SS *Great Eastern* between Foilhommerum on Valentia Island off the western coast of Ireland and Heart's Content, in eastern Newfoundland. Additional cables were laid between these locations in 1873, 1874, 1880, and 1894. The *Challenger* was looking for a new route.

While measuring the depth of the ocean, the expedition scientists discovered mountains on the ocean floor near the center of the Atlantic. When they investigated the floor to the north and south, they found that the mountains formed a chain extending as far as they sailed in either direction. They had discovered what is now called the Mid-Atlantic Ridge, running all the way from a point to the northeast of Greenland at 87° N to Bouvet Island, not far from the Antarctic Circle, at 54° S. The ridge crosses Iceland, where it is visible on land. In addition to Iceland, several high mountains along the ridge protrude above the surface as islands. These include Jan Mayen Island (Norway), the Azores (Portugal), St. Paul's Rock (Brazil), Ascension Island (United Kingdom), St. Helena (United Kingdom), Tristan da

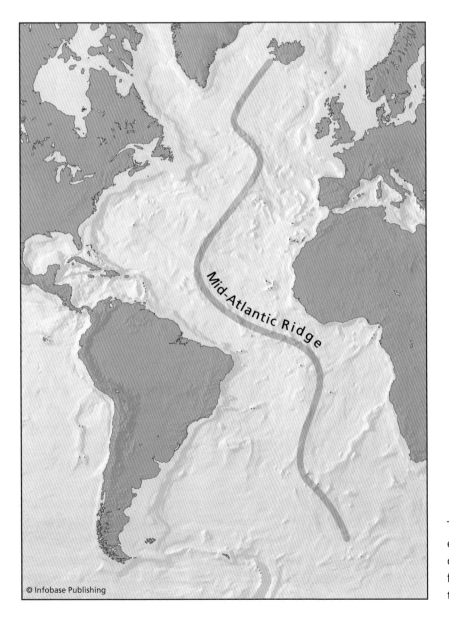

© Infobase Publishing

The Mid-Atlantic Ridge extends down the middle of the Atlantic Ocean, from Iceland all the way to Antarctica.

Cunha (United Kingdom), Gough Island (United Kingdom), and Bouvet Island (Norway). The map shows its location. Bermuda was once an island of the Mid-Atlantic Ridge, but seafloor spreading has carried it far to the side of the ocean (see "Harry Hess, Robert Dietz, and Seafloor Spreading" on pages 68–72).

The discovery was not altogether unexpected. The American oceanographer Matthew Fontaine Maury (1806–73; see "Matthew Fontaine Maury, Ocean Currents, and International Cooperation" on pages 29–33) had inferred its existence as early as 1850. Using only depth soundings, however, the *Challenger* scientists were unable to chart the ridge. They could only record their discovery of a chain of submarine mountains, with an estimate of their height. It was not until 1925 that the first attempt was made to map the ridge, using *sonar*—echo sounding used to detect underwater objects. Since then the ridge has been mapped in detail. It is 620–930 miles (1,000–1,500 km) wide and rises approximately 1.8 miles (3 km) above the ocean floor.

Since the discovery of the Mid-Atlantic Ridge other similar mid-ocean ridges have been discovered, and geologists now know that these submarine mountain chains are found in every ocean. Together they form a continuous range of mountains that is 40,400 miles (65,000 km) long. Other ridges, some on the margins of oceans, are not immediately joined to the continuous range, and when these are added the total length amounts to 49,700 miles (80,000 km). The ridges are 1.2–1.8 miles (2–3 km) high and 620–2,500 miles (1,000–4,000 km) wide, rising from ocean basins averaging three miles (5 km) in depth. They are truly immense and truly spectacular, and the fact that they remained undiscovered for so long illustrates the extent to which the oceans conceal their secrets. Most of their peaks are about 8,000 feet (2,440 m) below the surface and shrouded in total and perpetual darkness.

The Mid-Atlantic Ridge is a typical mid-ocean ridge. It sits above the Mid-Atlantic Rise, which is a bulge running the length of the ocean and that is thought to be caused by upward pressure in the *asthenosphere*—the weak zone in the upper *mantle*, where the rocks are slightly plastic and can deform. There is a rift valley along the center of the ridge for most of its length. The ridge is broken into sections by deep trenches that cross it, approximately at right angles. One of these, the Romanche Trench (also called the

Romanche Furrow and the Romanche Gap) extends for 186 miles (300 km) from 2° N to 2° S and 16° W to 20° W, dividing the ridge into northern and southern sections. The bottom of the Romanche Trench lies 25,460 feet (7,760 m) below sea level, making it one of the deepest places in the Atlantic Ocean. Water flows through the trench from west to east, and the water temperature at the bottom is 34.8°F (1.57°C).

The Romanche is not the deepest mid-Atlantic trench, however. The Puerto Rico Trench, located at the boundary between the Caribbean Sea and the Atlantic Ocean and slightly to the north of the island of Puerto Rico, is 28,232 feet (8,605 m) deep at its deepest point. The deepest part of the South Sandwich Trench, 60 miles (100 km) to the east of the South Sandwich Islands, is 27,652 feet (8,428 m) below the surface. It is called Meteor Deep, located at 55.67° S, 25.92° W. The South Sandwich Trench is 600 miles (965 km) long.

As sonar images improved, they revealed just how rugged the *bathymetry*—the seafloor equivalent of topography—of the Mid-Atlantic Ridge is. In the illustration the false colors are used to denote height above the ocean floor. The deepest regions are dark blue, and with increasing height above the floor the colors are pale blue, green, yellow, red, and white. The almost continuous blue section running from top to bottom is the rift at the center of the ridge. High mountain peaks line it on either side. In addition to the trenches, the ridge is broken by *faults*—a rock fracture caused when stresses on either side of a weakness exceed the strength of the rock and it breaks, the two sections

This sonar image shows the Mid-Atlantic Ridge running down the center of the picture, with one transform fault near the bottom and another large one at the top. The colors indicate depth, with dark blue the deepest, and pale blue, green, yellow, red, and white representing progressively shallower depths. *(Dr. Ken Macdonald, Science Photo Library)*

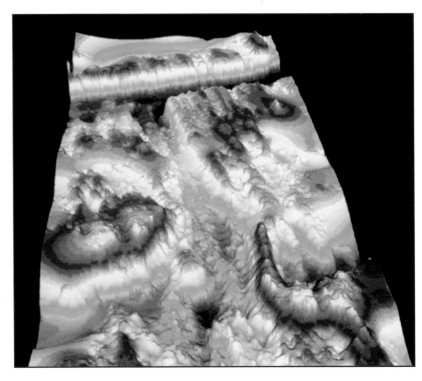

Ocean ridge

© Infobase Publishing

Transform faults occur along all mid-ocean ridges, approximately at right angles to the line of the ridge. The blocks move in a direction opposed to that occurring in otherwise similar faults on land, and it is this reversal, or trans-formation, that gives these faults their name.

moving in relation to each other. The image shows two faults, one near the bottom of the picture and another, deeper fault across the top.

These faults, running across the line of the ridge, occur along all mid-ocean ridges. The blocks on either side of the fault are displaced horizontally, but in the opposite direction from that found in other-wise similar faults on land (called *strike-slip faults*). Because the direction of movement is reversed, or transformed, these faults are known as *transform faults.* As the draw-ing shows, a succession of transform faults gives a mid-ocean ridge a zigzag shape.

During the 1950s studies of the Mid-Atlantic Ridge and the ridges found on the floors of other oceans revealed that all mid-ocean ridges are *seismically active*—regions of volcanic and earthquake activity (see "William Maurice Ewing, Mapping the Ocean Floor" on pages 62–65). Little by little these studies of the ocean floor led to a much wider understanding of the structure of the Earth's crust and the way continents move and oceans appear, widen, and eventually disappear.

WILLIAM MAURICE EWING, MAPPING THE OCEAN FLOOR

There is yet another way for scientists to study the ocean floor: They can generate shock waves resembling very small earthquakes and observe their effects. An earthquake occurs when rocks inside the Earth's crust are subjected to stresses that exceed their strength. The rocks fracture and move jerkily, usually in a series of jerks producing a main earthquake followed by smaller aftershocks. The movement of the rocks generates several types of shock waves, similar to sound waves, which propagate through the rocks. It is these waves that cause the surface rocks to move, producing the earthquake. They are known as *seismic waves,* from the Greek *seismos,* meaning earthquake, and the study of earthquakes is called *seismology. Seismometers* are instruments that detect seismic waves, and the waves are recorded on charts, called *seismographs.*

Seismographs allow seismologists to locate the source of an earthquake and its magnitude, but the study of seismic waves reveals much more than that. The speed that a shock wave travels depends on the density of the medium through which it travels. Sound travels through seawater at 3,232 mph (1,445 m/s), for example, but at only about 740 mph (331 m/s) through air at sea level pressure, and at about 670 mph (299 m/s) at about 40,000 feet (12,000 m). This is because seawater is much denser than air and air density decreases with increasing altitude. Shock waves travel still faster through solid rock, but their speed varies as they move through rocks of different densities. When a wave crosses the boundary between media with different densities, its speed changes, and it also changes direction. This is *refraction,* and by measuring it scientists can determine the density of the rocks through which the waves have passed and often the composition of the rocks. Their studies of seismic waves have greatly helped geologists to discover the structure of the solid Earth.

Geophysicists study the physical structure and properties of the Earth, and they use seismic waves as a tool. They are not restricted to the natural seismic waves resulting from earthquakes, however. They can manufacture their own very small earthquakes by detonating an explosion, dropping a very heavy weight, or releasing a sudden pulse of air from a device known as an air gun.

William Maurice Ewing (1906–74) was an American geophysicist who used *seismic reflection* and refraction, along with every other technique available to him, to study the ocean floor. In 1956 he was able to show that the Mid-Atlantic Ridge (see "Mid-Atlantic Ridge" on pages 58–62) continues around the southern tip of Africa and into the Indian Ocean and around South America into the Pacific Ocean.

He also studied the deep canyons in the seafloor that occur close to continents. The canyons look like extensions of river valleys, and some of them are large. The Congo Canyon, from the mouth of the

Congo River, is 500 miles (800 km) long and up to 3,900 feet (1,200 m) deep. The Hudson Canyon, from the mouth of the Hudson River, New York, extends for about 460 miles (740 km). These were thought to be the valleys of large rivers that had flowed when the sea level was much lower, perhaps during ice ages when large amounts of water were held as thick ice sheets on land. Ewing showed, in 1952, that this was not so. Turbulent underwater flows of sediment, set in motion by earthquakes or other disturbances and sliding down the slope of the continental slope like submarine avalanches, carve out the canyons.

During World War II Ewing discovered the *deep sound channel,* also called the SOFAR (*sound fixing and ranging*) *channel.* The channel was also discovered independently at about the same time by the Russian physicist Leonid Maksimovich Brekhovskikh (1917–2005). This is a horizontal ocean layer, close to the surface in latitudes higher than 60° N and 60° S descending to about 0.6 mile (1 km) below the surface in the Tropics, where the speed of sound is at a minimum. The channel acts as a waveguide, and low-frequency sound waves travel very long distances through it without dissipating. Militarily, it is useful for communicating with submarines, but whale calls also travel through it, and some biologists suspect humpback whales may swim into the channel deliberately to "sing" in order to communicate with distant whales.

William Maurice Ewing was born on May 12, 1906, at Lockney, Texas, the eldest of the seven surviving children (three died in infancy) of Floyd Ford Ewing, who was a hardware merchant and farmer, and Hope Hamilton Ewing. He won a scholarship to the Rice Institute (now Rice University) in Houston, where he enrolled in 1923, supporting himself by working at night for various oil companies. He majored in electrical engineering, later changing to physics and mathematics. Ewing graduated in 1926 and earned a master's degree in 1927 and a doctorate in 1931. From 1929 to 1930, Ewing was an instructor in physics at the University of Pittsburgh, moving in 1930 to Lehigh University, Pennsylvania, as an instructor in physics. He became an assistant professor in physics in 1936 and associate professor of geology in 1940. He participated in surveys of the continental shelf, and in September 1940 Ewing obtained permission from Lehigh to move his research group to the Woods Hole Oceanographic Institution in Massachusetts, where he became leading physicist in charge of developing underwater photography and sound technologies as part of the war effort. From 1944 to 1947, Ewing was associate professor at Columbia University. He was promoted to full professor in 1947, and in 1959 he was appointed Higgins Professor of Geology, a position he held until 1972.

Ewing's success as a research scientist persuaded the trustees of Columbia University to establish a geological observatory at Torrey Cliff, at Palisades, New York, formerly the weekend residence of the banker Thomas W. Lamont (1870–1948), which Lamont's widow, Florence Haskell Lamont (born Corliss, 1873–1952), had donated to the university after her husband's death. The Lamont Geological Observatory opened in 1949, with Maurice Ewing as its first director. It was renamed the Lamont-Doherty Earth Observatory (LDEO) in 1969, following the university's receipt of a gift from the Henry L. and Grace Doherty Charitable Foundation. The LDEO research vessel *Maurice Ewing* was named in his honor.

Maurice Ewing, usually known as "Doc" by those working with him, received many honors. He was elected to the National Academy of Sciences in 1948, the American Academy of Arts and Sciences in 1951, and the American Philosophical Society in 1959. He was vice president of the Geological Society of America (1952–55) and president of the Seismological Society of America (1955–57). He was named as a foreign member of the Geological Society of London (1964) and the Royal Society (1972). On April 28, 1974, Maurice Ewing suffered a major cerebral hemorrhage. He died at Galveston, Texas, on May 4, 1974.

MARIANA TRENCH

On March 23, 1875, sailors on board HMS *Challenger* measured an ocean depth of 4,475 fathoms (26,850 feet; 8,189 m)—by paying out a weighted rope more than five miles (8 km) long. In 1951 another Royal Navy survey ship, also called HMS *Challenger,* surveyed the site, this time using echo sounding, and gave it its name: the Challenger Deep. The map shows its location with respect to the largest nations of the region. The Challenger Deep, at 11.32° N, 142.25° E, is among the Mariana Islands, an archipelago of 15 islands, each of which is the summit of a volcano. Fais Island is 180 miles (289 km) to the southwest, and the island of Guam is 190 miles (306 km) to the northeast. The 1951 survey measured the depth at 5,960 fathoms (35,760 feet; 10,907 m).

Several other survey vessels have produced slightly different depth measurements, but no one questions the fact that the Challenger Deep is the deepest point on the surface of the Earth. It is more than 6,500 feet (2,000 m) deeper than Mount Everest is high.

Scientists have visited the Challenger Deep. *Trieste,* the U.S. Navy's manned *bathyscaphe,* descended to the bottom in 1960 (see "The *Trieste* and Its Voyage to the Challenger Deep" on pages 119–122). The Japanese unmanned submersible *Kaikō* reached the bottom in 1996—contact with *Kaikō* was lost during an expedition in 2003, and the vessel has never been recovered. In 2008 the Woods Hole Oceanographic Institution plans to send its unmanned submersible *Nereus* to explore and chart the deep.

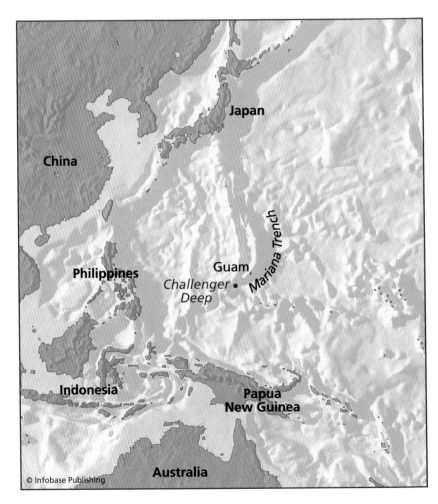

The Challenger Deep, on floor of the Mariana Trench, in the Mariana Islands, near the Pacific island of Guam, is the deepest point on the Earth's surface, about 36,000 feet (11,000 m) below sea level.

The Challenger Deep is more remote and much less accessible than the summit of Everest. Mountaineers can carry oxygen to help them breathe the thin air at high altitudes. This is inconvenient and tiring, but it is practicable. It is far more difficult to visit the Challenger Deep, where the problem is not the lack of air but the immense pressure. The weight of the overlying ocean means the pressure at the bottom is 15,750 pounds per square inch (108.6 megapascal, MPa); average sea-level atmospheric pressure is 15 pounds per square inch (0.101 MPa). No diving suit could withstand that pressure. Explorers have to remain inside the thick steel walls of their vehicle.

Mountaineers can enjoy magnificent views, but explorers of the deep can see only as far as the beams from their searchlights are able to reach. No daylight penetrates below a depth of about 650 feet (200 m), so the Mariana Trench is in utter darkness. Over most of its area it is also cold, with a water temperature that is always 34°F–39°F (1°C–4°C). Where it is not cold, the water is very hot. There are many *hydrothermal vents,* where emerging water is at an average temperature of about 570°F (300°C). Some vents discharge water rich in iron, copper, and manganese. This water is very hot—about 660°F (350°C)—and black, and the vents are called *black smokers.* Other vents flow more slowly, discharging cooler water rich in arsenic and zinc, which colors the water white. These are *white smokers.*

The bottom of the Mariana Trench is one of the most inhospitable places on the planet, yet there are organisms that thrive there. Some distance from the center of the hydrothermal vents, where the temperature is 105°F (40°C) or lower, there are bacteria that obtain energy by oxidizing sulfides dissolved in the vent fluid and that synthesize organic compounds. These bacteria provide a food base for a range of other animals. There are mussels, giant clams, crabs, tubeworms, and many other animals that belong to species different from those found in shallow water but are related to them. The most interesting of these are the beard worms belonging to the phylum Pogonophora, which have no digestive tract—no mouth, stomach, or gut—and obtain nourishment by absorbing nutrients released by bacteria. On March 2, 1996, the *Kaikō* collected a sample of mud from the floor of the Challenger Deep and returned it to the surface without contaminating it. A little of the sample was mixed with water, and when it was cultured, it was found to contain thousands of single-celled organisms, including many *alkaliphiles*—organisms that thrive in very alkaline conditions—and *thermophiles*—organisms that thrive at high temperatures and cannot tolerate cold (see the sidebar "Extremophiles" on page 142).

When the floor of the Mariana Trench was first surveyed, in 1951, the trench simply existed. No one knew how it came to exist or whether it was a permanent feature of the ocean floor. Similarly, mid-ocean ridges were regarded as submarine mountain ranges. No one knew whether these mountains were raised by the same processes as those that produced more familiar mountain ranges on land, such as the Rocky Mountains, Andes, and the Alps of central Europe. The first step toward finding an answer to these questions came with the discovery that the seafloor moves.

HARRY HESS, ROBERT DIETZ, AND SEAFLOOR SPREADING

By the 1950s scientists knew that the Mid-Atlantic Ridge was just one part of a system of mid-ocean ridges that runs along the floor of every ocean. They also knew that the Mariana Trench was just one of many similar trenches, albeit the deepest. Originally, the trenches were known as deeps, such as the Challenger Deep, but geologists began calling them trenches in the 1920s because of their resemblance to the long, narrow, deep trenches that became tragically familiar during World War I.

Oceanographers charted the trenches, but the first clue to their structure and the processes that formed them was discovered between 1923 and 1929 by a Dutch geophysicist, Felix Andries Vening Meinesz (1887–1966). Vening Meinesz studied gravity—the gravitational acceleration due to the force exerted by the Earth. In 1915, as a graduate student at the University of Delft, he wrote his dissertation on the subject of *gravimeters*—instruments for measuring gravitational acceleration. More particularly, he criticized the inaccuracy and inefficiency of the gravimeters in use at the time. Having completed his dissertation, Vening Meinesz, who was then working for the Dutch Gravitational Survey, set about designing an improved gravimeter. This was much more sensitive than those it replaced, and after he had measured gravity and gravitational anomalies throughout the Netherlands, his new instrument allowed Vening Meinesz to measure changes in gravity across the ocean floor. He made descents in small submersibles—which was something of a problem for him, for he was more than six feet five inches (2 m) tall. His gravimeter used light beams and mirrors to detect variations in the amplitude of two pendulums, which were recorded on film. His aim was to determine the precise shape of the Earth, but his measurements also revealed that oceanic trenches were sites where relatively dense rock was accumulating beneath lighter rock.

Interest in the ocean floor intensified during World War II as navies sought ways to detect submarines and to avoid such detection by hiding on the bottom. When the war began Harry Hammond Hess (1906–69) was a geologist on the faculty of Princeton University. He joined the navy, and after a time he became the captain of the USS *Cape Johnson,* a transport ship working in the Pacific that was fitted with sonar to help it avoid enemy submarines. Eventually Hess rose to the rank of rear admiral in the naval reserve, but while serving on the *Cape Johnson* he took the opportunity of using the ship's sonar to explore the ocean floor. As the ship sailed among the war zones of the Pacific, Hess surveyed the floor around the Mariana Islands, the Philippines, and Iwo Jima, an island to the north of the Marianas and part of the Japanese island system. He discovered flat-topped *seamounts*—islands rising from the ocean floor that do not reach the surface—that he called *guyots* in honor of the Swiss-American geologist Arnold Henry Guyot (1807–84).

Hess also studied mid-ocean ridges, and in 1960 he wrote a report to the Office of Naval Research that was published in 1962 as *History of Ocean Basins.* In this work Hess set out a radical new theory. He proposed that mid-ocean ridges are places where magma rises through the crust. The molten rock cools and solidifies to form the mountains of the ridges, but the rift that allows the magma to emerge is a split in the oceanic crust. The crust to either side of the rift is slowly moving away from the center, and the seafloor is spreading, the magma forming new oceanic crust to fill the gap. Obviously, if the seafloor is spreading there must also be places where oceanic crust is being removed, and there is: the ocean trenches. Hess had worked with Vening Meinesz, and he recognized the significance of the Dutch scientist's gravity studies of trenches. The diagram illustrates the process that Hess helped elucidate. Oceanic *lithosphere*—the crust together with the brittle uppermost layer of the mantle—is being pulled downward into the asthenosphere. The process is called

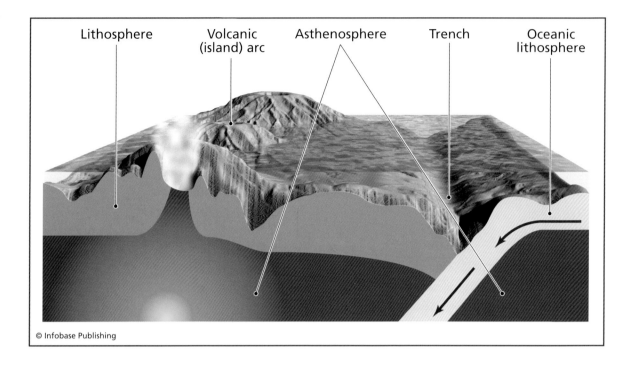

Lithosphere   Volcanic   Asthenosphere   Trench   Oceanic
               (island) arc   lithosphere

© Infobase Publishing

An oceanic trench occurs in a subduction zone where oceanic crust—the ocean floor and the sediment on its surface—is sinking into the asthenosphere and from there into the Earth's mantle. A volcanic island arc develops in the lithosphere ahead of the trench, from volcanic activity caused by the disturbance.

subduction, and the region where it occurs is a *subduction zone.* This produces a trench above the point where the oceanic crust is being subducted, and ahead of the trench there is a seismically active region where volcanoes produce an arc of islands approximately parallel to the subduction zone and trench.

Harry Hammond Hess was born on May 24, 1906, in New York City. He entered Yale University in 1923 to study electrical engineering, but after two years he changed to geology, in which he graduated in 1927. Hess then left Yale and went to work for two years as an exploration geologist in Southern Rhodesia (now Zimbabwe) before entering Princeton University to commence his graduate studies. He received his doctorate in 1932 and then taught for a year at Rutgers University, followed by a year spent as a research associate at the Geophysical Laboratory in Washington, D.C. In 1934 Hess joined the faculty of Princeton University, where he remained for the rest of his career, apart from his wartime naval service. He was also a visiting professor at the University of Cape Town, South Africa, from 1949 until 1950 and at the University of Cambridge, England, in 1965.

Hess was elected to the National Academy of Sciences in 1952, the American Philosophical Society in 1960, and the American Academy of Arts and Sciences in 1968. At different times he served as president of two sections of the American Geophysical Union. On August 25, 1969, he died from a heart attack while chairing a meeting of the Space Science Board of the National Academy of Sciences.

It was not Harry Hess who coined the term seafloor spreading, however, but another American geophysicist and oceanographer, Robert Sinclair Dietz (1914–95), who reached the same conclusion as Hess but independently of Hess's work. Robert Dietz was born on September 14, 1914, in Westfield, New Jersey. His father, Louis Dietz, was a civil engineer. Robert entered the University of Illinois in 1933 to study geology, where he obtained his bachelor's and master's degrees and, in 1941, his Ph.D. After graduating he was drafted into the Army Air Corps and served as a pilot stationed in South America. After the war he remained in the reserve for 15 years, rising to the rank of lieutenant colonel.

Dietz did much of his doctoral work at the Scripps Institution of Oceanography, which is part of the University of California at San Diego, and when his military service ended a colleague from Scripps invited him to organize a group at the Naval Electronics Laboratory in San Diego to study the seafloor. While he was at the Naval Electronics Laboratory, Dietz sailed on several oceanographic cruises exploring the Pacific Basin.

Soon after they came onto the market, Dietz's group bought Aqua-Lungs, which were developed in 1943 by the French engineer Émile Gagnan (1900–79) and the French naval officer, underwater explorer, and later filmmaker Jacques-Yves Cousteau (1910–97), and learned how to use them. Dietz had met Cousteau during his time in London, and Cousteau had introduced him to another pioneer of underwater exploration, Jacques Piccard (1922–2008). Piccard had already designed a submersible, and Dietz offered to seek support from the Office of Naval Research, where he was then working, for the construction of the bathyscaphe *Trieste* (see "The *Trieste* and Its Voyage to the Challenger Deep" on pages 119–122). In subsequent years Dietz made many dives along the coasts of California and Baja California. He and his colleagues mapped and photographed submarine canyons.

Dietz remained at the Naval Electronics Laboratory from 1946 until 1963, and from 1950 until 1963 he was an adjunct professor at the Scripps Institution. His time at the Laboratory was interrupted in 1953 when he was a Fulbright Scholar at the University of Tokyo and again from 1954–58 when he worked at the Office of Naval Research, based in London. Dietz joined the U.S. Coast and Geodetic Survey, in Washington, D.C., in 1963. The survey later moved its offices to Miami, becoming the Environmental Sciences Administration, which was absorbed into the National Oceanic and Atmospheric Administration (NOAA) in 1970. Dietz retired from NOAA in 1975 and held visiting professorships at several universities before accepting a tenured position at the University of Arizona in 1977. In 1985 he became an emeritus professor.

As well as his research on seafloor spreading and continental drift, Robert Dietz was also interested in lunar craters and craters on Earth that had been caused by impacting bodies. He died from a heart attack at his home in Tempe, Arizona, on May 19, 1995.

TETHYS, PANTHALASSA, PANGAEA, AND THE DRIFTING CONTINENTS

When reasonably accurate maps of the world first became available in the 16th century, they revealed a curious fact. If the Atlantic Ocean could somehow be removed, it looked as though the continent of Africa might fit quite snugly into the Americas. This might be pure coincidence, of course, but it intrigued many geographers. Over the following centuries scientific explorers discovered that the apparent geographic fit extended beyond the shapes of coastlines. The rocks on either side of the Atlantic were so similar, they might once have been joined. There were species of plants and animals that occurred on lands separated by vast oceans their ancestors could not possibly have crossed. There was also the fact that there were fossilized seashells in the rocks found on many mountains, high above sea level.

The Austrian geologist Eduard Suess (see "Osmond Fisher and the Origin of the Pacific Basin" on pages 40–43) was especially interested in the rocks of the Alps, and his studies led him to conclude that the rocks comprising these mountains had once lain on the bottom of the sea. He believed that the Mediterranean Sea was all that remained of what in 1893 he called the *Tethys* Ocean. Suess also noted that fossils of plants with long, tongue-shaped leaves were found in South America, Africa, and India (they are also found in Antarctica, but Suess did not know this). Together, these long extinct plants are known as the *Glossopteris* flora, after one of the most prominent genera. Suess suggested that the best way to explain this distribution would be to

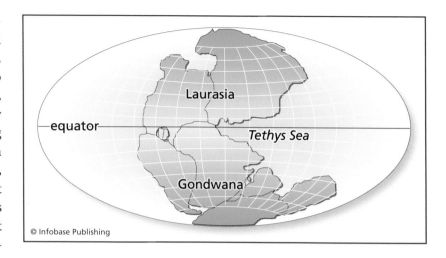

© Infobase Publishing

suppose that the three continents had once been joined into a supercontinent he called Gondwanaland—it is now known as *Gondwana*. Suess summarized his ideas in *Das Antlitz der Erde,* a book that was published between 1885 and 1901 and that appeared in an English translation between 1904 and 1925 as *The Face of the Earth.*

Eduard Suess lived before the development of modern plate tectonic theory (see "Plates, Ridges, and Trenches" on pages 79–84), but geologists now recognize that his theory was essentially correct. The southern continents were once joined, forming the supercontinent Gondwana, and the northern continents were also joined to form another supercontinent called *Laurasia.* The Tethys Sea, representing part of Suess's Tethys Ocean, separated the two supercontinents on their eastern side. The illustration shows a map of the world as geologists believe it appeared about 200 million years ago. Laurasia and Gondwana have not quite separated, and Gondwana is beginning to break apart. There is a narrow sea between South America and Africa, and Antarctica and India have broken away.

It was not long after Eduard Suess proposed the existence of Gondwana that Alfred Wegener (1880–1930), a German meteorologist, carried the idea a stage further. The map of Gondwana and Laurasia shows the two supercontinents linked by a narrow isthmus. Wegener suggested that they had been joined much more closely. He assembled evidence from rock formations, fossils, and the distribution of plants and animals to construct a map in which all of the

Two hundred million years ago the northern continents of North America and Eurasia were joined, forming the supercontinent Laurasia. To the south, the supercontinent of Gondwana was starting to break apart.

present continents were joined, forming a single supercontinent that he called *Pangaea.* The Greek *pan* means all, and *gaea* means Earth, so Pangaea means "all of the Earth." The supercontinent was shaped rather like the letter C, and it extended from far northern latitudes almost to the South Pole. The map shows its general shape and the positions of the modern continents within it. Pangaea was surrounded by a worldwide ocean called *Panthalassa* or the Panthalassic Ocean—the name means "all of the ocean"—and the large indentation on the eastern side of the supercontinent was called the Tethys Sea. Pangaea formed about 265 million years ago, during the late Permian period (see the sidebar "Geologic Timescale"), and began to break apart in the middle Jurassic period, about 175 million years ago, so Pangaea existed for about 90 million years.

The fact that Pangaea formed when masses of land moved together and that the supercontinent later broke apart suggests that the continents are able to alter their positions, and that was Wegener's central point. He called the process "continental displacement" *(Verschiebung der Kontinente).* It is now known as *continental drift.* Geologists have since discovered that there were earlier supercontinents. Each in its turn formed and later broke apart. *Pannotia* was the supercontinent that preceded Pangaea; it existed in the Neoproterozoic era, 1,000–542 million years ago. Earlier still the supercontinent *Rodinia* existed in the Mesoproterozoic era, 1,600–1,000 million years ago.

Alfred Lothar Wegener was born on November 1, 1880, in Berlin. His father was a minister and director of an orphanage. Alfred studied at the universities of Heidelberg, Innsbruck, and Berlin, and in 1904 the University of Berlin awarded him a Ph.D. in planetary astronomy. Having qualified, Wegener immediately changed direction. In 1905 he accepted a job at the Royal Prussian Aeronautical Observatory, where he used kites and balloons to study the upper atmosphere. He also used hot-air balloons, and in 1906 he and his brother Kurt remained aloft for more than 52 hours, establishing a world endurance record. Also in 1906 Alfred served as the official meteorologist on a Danish expedition to Greenland. Alfred Wegener went on to become a highly distinguished meteorologist. He became a lecturer in meteorology and astronomy at the University of Marburg in 1909. In 1912 he married Else Köppen, daughter of Wladimir Köppen (1846–1940), the most eminent climatologist

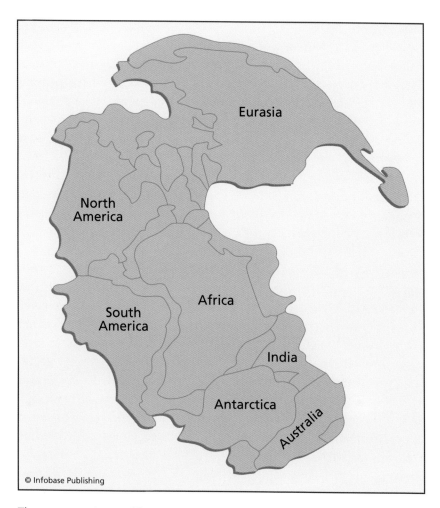

© Infobase Publishing

The supercontinent of Pangaea was approximately the shape of a letter C, with the Tethys Sea occupying the deep indentation on the eastern side. The surrounding ocean is called Panthalassa. Geologists have used geological structures and assemblages of fossils to identify the places where the modern continents were joined and, therefore, their positions within Pangaea.

in Germany, and Wegener collaborated with his father-in-law on a number of projects.

Meteorology was Wegener's primary interest, but it was not his only one. Intrigued by the apparent fit of the coastlines on either side of the Atlantic Ocean, he studied the scientific papers that had been published on the topic up to that time. Many of these, rejecting

Suess's theory, proposed the existence of a land bridge that had once extended across the Atlantic, allowing animals to migrate between Africa and South America. The land bridge had later disappeared. Wegener disagreed and developed his own theory to explain what he believed had happened, assembling a large amount of evidence to support the theory. In 1912 he outlined his idea in a short book entitled *Die Entstehung der Kontinente und Ozeane* (The origin of the continents and oceans). His theory was that the continents had once been joined together and that they had moved apart, eventually reaching their present positions. The book aroused little interest, and Wegener returned to Greenland as part of a four-man expedition, from which he returned in 1913.

War broke out on August 1, 1914, and before long Wegener was drafted into the German army. He was wounded early in the war, and during a long period of recuperation he elaborated his idea and wrote an expanded version of his 1912 book, keeping the same title. This was published in 1915, but because of the war and the fact that Wegener wrote in German, few scientists outside Germany and Austria read it. It was not until the 1920s that the third edition of the book was translated into other languages—in English as *The Origin of the Continents and Oceans*, published in 1924. Eventually six editions were published, but at first its reception was extremely hostile. Geologists derided it, and many of their attacks were directed at Wegener in person. He had demolished the idea of land bridges by pointing out that continental rocks are of a different and less dense type than the rocks that comprise the ocean floors, but he could suggest no mechanism by which objects the size of continents could move about the Earth, and that was his principal fault. He also seriously miscalculated the rate at which this movement occurs. It was not until the discovery of seafloor spreading in the 1950s (see "Harry Hess, Robert Dietz, and Seafloor Spreading" on pages 68–72) that opinions began to change and Wegener's reputation recovered. His theory has now joined the theory of seafloor spreading within the larger theory of plate tectonics (see "Plates, Ridges, and Trenches" on pages 79–84). Wegener did not benefit from being recognized as a visionary, however, because by then he had been dead for more than 20 years.

After recovering from his war wounds, Wegener spent the rest of the war working in the military meteorology service. When the war ended he returned to his old post at Marburg, but hostility from other

THE GEOLOGIC TIMESCALE					
EON/ EONOTHEM	ERA/ERATHEM	SUB-ERA	PERIOD/ SYSTEM	EPOCH/ SERIES	BEGAN MA
Phanerozoic		Quaternary	Pleistogene	Holocene	0.11
				Pleistocene	1.81
	Cenozoic	Tertiary	Neogene	Pliocene	5.3
				Miocene	23.3
			Paleogene	Oligocene	33.9
				Eocene	55.8
				Paleocene	65.5
	Mesozoic		Cretaceous	Late	99.6
				Early	145.5
			Jurassic	Late	161.2
				Middle	175.6
				Early	199.6
			Triassic	Late	228
				Middle	245
				Early	251
	Paleozoic	Upper	Permian	Late	260.4
				Middle	270.6
				Early	299
			Carboniferous	Pennsylvanian	318.1
				Mississipian	359.2
			Devonian	Late	385.3
				Middle	397.5
				Early	416
		Lower	Silurian	Late	422.9
				Early	443.7
			Ordovician	Late	460.9
				Middle	471.8
				Early	488.3
			Cambrian	Late	501
				Early	542
Proterozoic	Neoproterozoic		Ediacaran		600
			Cryogenian		850

(continues)

THE GEOLOGIC TIMESCALE *(continued)*					
EON/ EONOTHEM	ERA/ERATHEM	SUB-ERA	PERIOD/ SYSTEM	EPOCH/ SERIES	BEGAN MA
	Mesoproterozoic		Tonian		1,000
			Stenian		1,200
			Ectasian		1,400
			Calymmian		1,600
	Palaeoproterozoic		Statherian		1,800
			Orosirian		2,050
			Rhyacian		2,300
			Siderian		2,500
Archaean	Neoarchaean				2,800
	Mesoarchaean				3,200
	Palaeoarchaean				3,600
	Eoarchaean				3,800
Hadean	Swazian				3,900
	Basin Groups				4,000
	Cryptic				4,567.17

Source: International Union of Geological Sciences, 2004.

Note: Hadean is an informal name. The Hadean, Archaean, and Proterozoic Eons cover the time formerly known as the Precambrian. Tertiary had been abandoned as a formal name, and Quaternary is likely to be abandoned in the next few years, although both names are still widely used. Ma means "millions of years ago."

scientists meant no German university would offer him a professorship. In 1924 the University of Graz, Austria, appointed him professor of meteorology and geophysics. In 1930 he returned to Greenland for a third time, as leader of a team of 21 scientists and technicians on an expedition to study the climate over the ice cap. His body was discovered on the ice cap on May 12, 1930, and his colleagues assumed he had died from a heart attack. They built a mausoleum from ice blocks over the body and returned later to erect an iron cross 12 feet (3.7 m) tall at the spot. Both have long since disappeared beneath the accumulating ice.

PLATES, RIDGES, AND TRENCHES

No one could imagine how entire continents could move, but despite this serious objection Wegener's idea found a few supporters. One of the earliest was Alexander Logie Du Toit (1878–1948), a South African geologist who worked for the Geological Commission of the Cape of Good Hope, producing geological maps of the Karoo, a region of semidesert located mainly in Cape Province but extending into southern Namibia. The map shows its location.

Du Toit was born in Newlands, a suburb of Cape Town, on March 14, 1878, and educated at the University of the Cape of Good Hope. He then moved to Scotland, graduating with a degree in mining engineering from the Royal Technical College, Glasgow (now the University of Strathclyde) in 1899. After studying geology in London at the Royal College of Science (now part of Imperial College), Du Toit returned to Glasgow as a lecturer in geology, mining, and surveying at both the University of Glasgow and the Royal Technical College. It was in 1903 that his appointment to the Geological Commission of the Cape of Good Hope took him back to South Africa. He died at Cape Town on February 25, 1948.

In 1923 Du Toit received a grant from the Carnegie Institution of Washington, which he used to study the geology of Brazil, Paraguay, and Argentina. He found the rock formations of South America so similar to those of South Africa that he became convinced the two continents had once been joined. His discoveries supported Wegener's theory. He described them in *A Geological Comparison of South America with South Africa,* a book published in 1927, and in 1937 he published a second, longer book, *Our Wandering Continents,* in which he further developed the idea.

Alexander Du Toit was not the only scientist to support Wegener. The English geophysicist Arthur Holmes (1890–1965) proved Wegener's strongest supporter. Holmes was born at Hebburn-on-Tyne in the northeast of England, on January 14, 1890. At 17 he won a scholarship to study physics at the Royal College of Science (now Imperial College) in London. In his second year he switched to geology, graduating in that subject in 1910, but he retained his strong interest in physics.

The French physicist (Antoine-) Henri Becquerel (1852–1908) discovered radioactivity in 1896, triggering a huge and rapid expansion of research into this phenomenon. Physicists found that by emitting

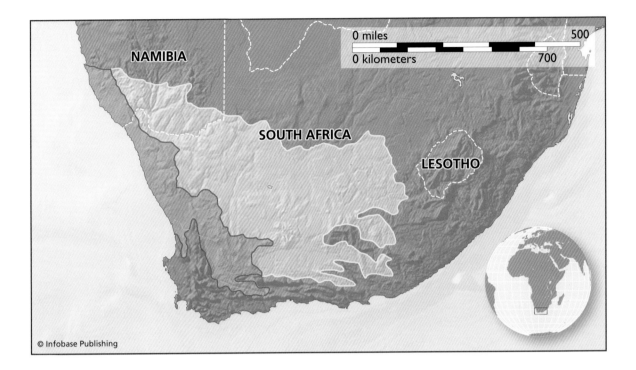

© Infobase Publishing

The Karoo is a region located mainly in South Africa's Cape Province but extends into southern Namibia. The yellow line marks its boundaries. Ecologically, there are two parts to the Karoo, separated by the red line. The western section, nearer the coast, supports more vegetation than the drier eastern section.

radiation, radioactive elements decay to nonradioactive, stable elements at a very precise and, therefore, predictable rate. In 1904 the New Zealand–British physicist Ernest Rutherford (1871–1937) became the first person to determine the age of a rock using this method, now called *radiometric dating*.

Arthur Holmes began applying the new technique, and in 1911 a friend read a paper he had written to members of the Royal Society, describing how Holmes had used the rate of the radioactive decay of uranium to lead to measure the age of a rock from Norway—it was 370 million years old. Holmes was unable to read the paper himself because by that time he was in Mozambique prospecting for minerals, having taken a job to earn some money. The scholarship keeping him at university paid only £60 a year, worth about £4,400 ($8,800) in today's money. No valuable minerals were found after six months, and Holmes caught blackwater fever and malaria. He was so ill that an obituary notice was telegraphed to England, but he recovered and immediately sailed for home. His impaired health meant he was not drafted into the military in World War I.

In 1913 Arthur Holmes published his first book, *The Age of the Earth,* in which he proposed the first geologic timescale (see the sidebar, "Geologic Timescale," on page 77). Previous tables summarizing the history of the Earth's rocks had arranged them in a chronological succession and had divided the history into a sequence of named episodes—eons, eras, periods, and epochs—but they had not attached ages to any of these, or to the Earth itself. Holmes's calculations suggested that the Earth was about 1.6 billion years old (the age is now known to be about 4.56 billion years), and he suggested that the Cambrian period began about 600 million years ago; the date is now set at 542 million years ago. Most scientists at the time believed the Earth could not be more than 100 million years old, so Holmes's proposed age met stiff opposition. In 1913 Holmes was 23 years old and had not yet received his doctorate!

After graduating, Arthur Holmes became a demonstrator on the staff of Imperial College while continuing his studies. He obtained his Ph.D. in 1917. Needing to increase his income to support a wife and young son, in 1920 he accepted a position as chief geologist with an oil company in Burma (now called Myanmar), but the company failed to pay him, and in 1922 he was forced to return to England, penniless and mourning the loss of his son, who had died from dysentery. In 1924 he was offered the post of head of the newly formed geology department at the University of Durham, in the northeast of England. He remained at Durham until 1943, when he became regius professor of geology at the University of Edinburgh. Holmes published his second and possibly most famous book in 1944. It was entitled *Principles of Physical Geology,* and in it he revised his measurement of the Earth's age to 4.5 billion ± 100 million years. He retired in 1956 and died on September 20, 1965.

By the time Holmes returned from Burma in 1922 the controversy over the age of the Earth had quieted down. Almost all geologists accepted that the Earth was very ancient, and their arguments centered on a new controversy, over continental drift. Holmes supported the idea and proposed an explanation for it. In 1928 he suggested that the very high temperature generated by radioactive decay in the lowest region of the Earth's mantle produced convection currents. Hot mantle material rose, very slowly, through the dense but plastic rock, cooled beneath the base of the crust, and sank back into the mantle.

This gradual movement produced fissures in the oceanic crust and transported continents. Holmes continued teaching this to his students, despite the lack of support his idea received from other geologists, but eventually he was vindicated with the discovery of seafloor spreading (see "Harry Hess, Robert Dietz, and Seafloor Spreading" on pages 68–72). That was in the 1960s, a few years before he died.

In 1964 the Royal Society sponsored a symposium on continental drift. The Canadian geophysicist John Tuzo Wilson (1908–93) attended. Harry Hess's work had convinced Wilson of the reality of seafloor spreading, and Sir Edward Bullard (1907–80) presented further evidence at the meeting. Bullard, head of the department of geodesy and geophysics at the University of Cambridge, had formerly been professor of geophysics at the University of Toronto, director of the National Physical Laboratory, a British government institution near London, and since 1963 he was also a professor at the University of California. Bullard's huge reputation attracted talented scientists to work under him at Cambridge, among them Drummond Hoyle Matthews (1931–97), who was a research fellow, and Matthews's postdoctoral researcher Frederick John Vine (born 1939; since he retired in 1998 Vine has been an emeritus professor at the University of East Anglia).

In 1962 Matthews and Vine had surveyed part of the Indian Ocean mid-ocean ridge on the floor of the Gulf of Aden, where they found patterns of magnetic stripes running parallel to the ridge on either side of it. A compass needle placed over one stripe would point north, but placed over the adjacent stripe it would point south. Similar stripes had been discovered earlier running parallel to the Pacific ridge.

Atoms of certain elements, such as iron, align themselves with the Earth's magnetic field when they are below a critical temperature called the *Curie temperature*; this is 1,247°F (675°C) for hematite (the major iron ore) and 1,067°F (575°C) for magnetite (an iron oxide also called lodestone). As molten rock cools and its temperature falls below that threshold, the magnetic atoms it contains align themselves with the Earth's field, and when the rock solidifies they are locked in position. Consequently, rocks record the alignment of the Earth's magnetic field as it was at the time when they solidified. The Earth behaves rather like a bar magnet, but from time to time it reverses its polarity: The magnetic North Pole becomes the South

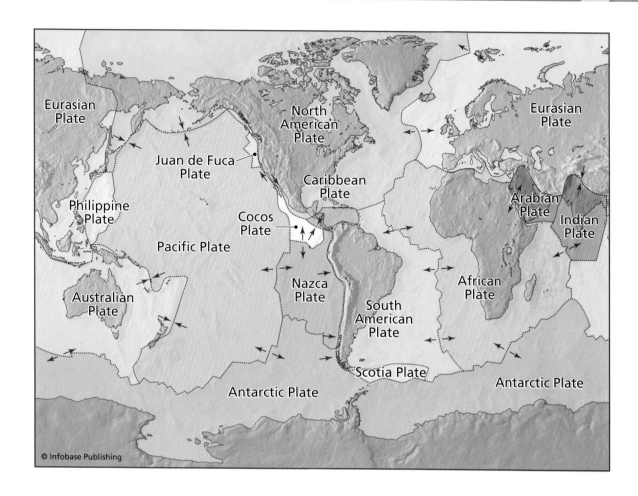

Eurasian Plate

North American Plate

Eurasian Plate

Juan de Fuca Plate

Caribbean Plate

Arabian Plate

Indian Plate

Philippine Plate

Cocos Plate

Pacific Plate

Nazca Plate

African Plate

Australian Plate

South American Plate

Scotia Plate

Antarctic Plate

Antarctic Plate

© Infobase Publishing

Pole and vice versa. The magnetic stripes running parallel to mid-ocean ridges record the history of these reversals. Matthews and Vine realized that the stripes proved that molten rock emerging from the ridge had solidified at different times. This proved the theory of seafloor spreading. In 1963 they published their findings as an article in *Nature*.

Early in 1965, Tuzo Wilson, Harry Hess, Drummond Matthews, and Fred Vine met at Cambridge to discuss seafloor spreading and its implications. Wilson published a paper in *Nature* suggesting that mountain belts, earthquake zones, and volcanoes form lines that end at transform faults and that the transform faults link the ridges, trenches, and mountain belts, dividing the Earth's crust into sections. Soon after that the American geophysicist William

The Earth's crust is broken into sections called plates, which move in relation to one another. The map shows the seven major plates (African, Antarctic, Eurasian, Indian, North American, Pacific, and South American) and several of the minor plates.

Jason Morgan (born 1935), at Princeton University, and the English geophysicist Dan McKenzie (born 1942), at Cambridge University, working independently of each other, both concluded that the crust is composed of separate blocks, or plates. Ridges are the boundaries between plates that are moving apart, and trenches are the boundaries between plates that are colliding and where one plate is being subducted beneath the other. Mountain ranges such as the Himalayas form where two continental plates collide—in this case the Indian plate is moving northward into the Eurasian plate. The theory of *plate tectonics* was born. It proposes that the rigid crust is divided into plates that move, carried by convection currents in the mantle.

Geophysicists now recognize seven major *tectonic plates*: the African, Antarctic, Eurasian, Indian, North American, Pacific, and South American. There are also minor plates such as the Philippine, Juan de Fuca, Nazca, Cocos, Caribbean, Arabian, and Scotia plates, as well as 14 microplates, including the Caroline, Burma, Tonga, and Galápagos microplates. The plates range in size from the Pacific plate, with an area of 39,874,000 square miles (103,300,000 km^2), to the Galápagos microplate, with an area of 4,632 square miles (12,000 km^2). The map shows the major and minor plates. The theory of plate tectonics now provides the foundation on which much of modern geology is built.

Measuring the Depth and Flow

Maps of the oceans depict coastlines and islands. They show the dimensions of the oceans. Navigators need more information than this, however. A navigator who fails to take account of the ocean currents may unwittingly direct a ship far from its planned course at a speed that may be faster or slower than predicted or, in the days of sail, than the winds might suggest. This matters, because supplies of food and water are limited. When the lookout sights land and the ship enters coastal waters, the navigator needs to know the depth of water and the location of reefs and sandbanks.

Scientists need other information, likely to be of little interest to sailors, about the water itself. Seawater is salty and undrinkable, but it supports living organisms that must find sustenance in it. So the water cannot be a simple salt solution. Chemists have analyzed seawater in great detail.

This chapter recounts some of those discoveries. It tells of the navigators and scientists who explored the substance of the oceans—the water itself—and the way ocean currents move. It also describes how one of the greatest physicists of the 19th century invented an ingenious device that accurately measured the depth of water beneath a moving ship.

FERDINAND MAGELLAN, MEASURING THE DEEP

On January 24, 1521, a sailor on board the *Trinidad*, commanded by Ferdinand Magellan (1480–1521), cast a sounding line over the side.

The line was 400 fathoms (2,400 feet; 730 m) long, and it failed to find the bottom. The ship was crossing the "Sea of the South"—the sea was so calm that Magellan renamed it the Pacific (peaceful) Ocean. There is a legend that Magellan rather illogically assumed the failure of the sounding line to reach the bottom meant they had arrived at the deepest point in the ocean. It is much more likely that the *Trinidad* was approaching the coral atoll of Puka Puka and the great navigator was searching for water shallow enough to drop anchor so a party could go ashore to find food and freshwater. The crew had almost exhausted its water supply, many of the sailors were suffering from scurvy, and their only food consisted of biscuits fouled by rats, and they were reduced to chewing leather. Puka Puka is one of the atolls of the Tuamotu Archipelago in French Polynesia.

Magellan had left Spain on September 20, 1519, with a fleet of five ships, the *Trinidad, San Antonio, Concepción, Victoria,* and *Santiago.* By the time he reached Puka Puka, the *San Antonio* had deserted the expedition, and the *Santiago* had been wrecked while taking depth soundings. The three remaining ships were badly damaged by teredos, a genus of shipworms, which are bivalve mollusks that feed on the planking of wooden ships.

The purpose of the expedition was political. Magellan was a Portuguese aristocrat and probably born in Oporto. His Portuguese name was Fernão de Magalhães. He had fought for Portugal to oppose Muslim power in the Indian Ocean, but in 1516 the Portuguese king released him from his service and, accompanied by the Portugese cosmographer Rui Faleiro, Magellan traveled to Spain, where he offered his services at the royal court at Valladolid, becoming a Spanish citizen, with the Spanish name Hernando de Magallanes.

Portugal and Spain had both claimed territory in the Americas, and Pope Alexander VI resolved their dispute in 1493 by ordering that all lands to the west and south of an agreed line should belong to Spain and lands to the east and south to Portugal. Magellan now sought to convince the Spanish king that the Moluccas, known then as the Spice Islands and the source of spices that commanded huge prices in Europe, lay on the Spanish side of the line. He said he could prove this by sailing westward from Spain and finding a strait—which he claimed to know existed—that would lead him across the papal line and to the islands. The king agreed, and the fleet was duly fitted out and sailed. They reached the Brazilian coast,

EL NIÑO

Usually the surface air pressure is high over the eastern South Pacific near Tahiti and low in the west, near Darwin, Australia. The difference in pressure intensifies the southeasterly trade winds blowing across the tropical South Pacific Ocean, and the trade winds drive a surface current. Tropical surface waters are warm, and the current carries warm water from east to west, away from South America and toward Indonesia, where the warm water collects in a deep pool. Evapora-

tion and convection from this warm pool give Indonesia a warm, wet climate. The layer of warm water off the South American coast is much shallower, and in many places cold water wells up to the surface, bringing nutrients carried from the Southern Ocean by the cold Peru Current, which flows northward parallel to the coast.

At intervals of about seven years this pattern weakens or reverses. The difference in pressure is
(continues)

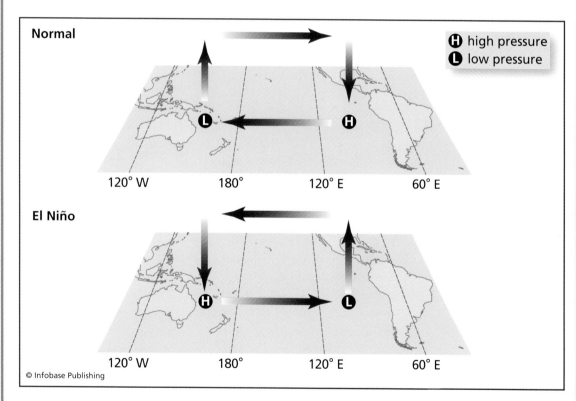

© Infobase Publishing

Normally, surface pressure is high in the eastern Pacific and low in the western Pacific. The difference in pressure intensifies the southeasterly trade winds, which in turn drive a surface sea current flowing from east to west. During an El Niño event, the pressure difference weakens, weakening the winds and current.

(continued)

reduced, weakening the trade winds, which may cease to blow. The east-west current slackens or in extreme cases reverses direction. Warm water drains from the warm pool, producing drier weather, and often drought, in Indonesia, and warm water accumulates off South America, suppressing the upwelling of nutrient-rich water and producing wet weather in western South America, where the climate is otherwise arid.

This change in the weather usually reaches a peak in December (midsummer in the Southern Hemisphere), and because it brought rain to water the crops, heralding an abundant harvest, Christian missionaries called it El Niño, the (boy) child, implying that it is a Christmas gift. El Niño events affect weather patterns over a much wider area, however, and the plentiful harvest for the farmers is offset by the disappearance of fish dependent on the nutrient-rich waters and drought in Indonesia and Australia.

Sometimes an El Niño is followed by its opposite, called La Niña, when the pressure pattern strengthens, deepening the warm pool and making the Indonesian weather even wetter and that of South America even drier. The periodic change in the pressure pattern is called a Southern Oscillation, so the full cycle is known as an ENSO (El Niño–Southern Oscillation) event. The diagram illustrates the normal and El Niño conditions.

then sailed southward along uncharted coasts through weather that steadily deteriorated, exploring every indentation and estuary in the hope of finding the strait. Some estuaries were very deep, especially that of the Río de la Plata. Along the way the sailors came across *patos sin alas*—ducks without wings—that they had to skin rather than pluck before cooking: They had encountered penguins. Far to the south, the ships spent most of the winter in a bay at what is now Puerto San Julián, Argentina. There some of the Spanish captains mutinied against their Portuguese commander. Magellan suppressed the rebellion, executing one of the leaders and abandoning another to his fate on shore. It was there, however, that they met a population of extremely large shepherds who herded guanacos—the sailors called them camels without humps—and dressed in guanaco skins. They called the men *patagones*—thought to mean *big feet*—and the land where they dwelt became Patagonia.

The fleet left San Julián on August 24, 1520, and on October 21 the ships entered what proved to be the strait Magellan had been seeking and which now bears his name. On leaving the strait they turned north but were swept along by the Peru (or Humboldt) Current, which held them close to the coast of what is now Chile. They turned westward in January and crossed the ocean, but instead of heading directly for the Moluccas, on March 6, 1521, they made landfall at Guam in the Mariana Islands, more than 930 miles (1,500 km) northeast of the Moluccas. Magellan was a highly skilled and very experienced navigator, so this diversion has puzzled historians. Scientists now suspect that Magellan may have heard there were food shortages in the Moluccas and decided to make for Guam. Those shortages, possibly due to drought, and the calm conditions that allowed the ships to drift northward on the Peru Current may both have been caused by an El Niño event (see the sidebar).

The expedition left Guam on March 9, bound for the Philippines to take on additional supplies before heading to the Moluccas. They arrived at Mazaua in the Philippines, but it was there, on April 27, that Magellan was killed in a fight with local inhabitants. Of the original fleet of five ships, only the *Trinidad* and *Victoria* reached the Moluccas and only the *Victoria*, commanded by the Basque navigator Juan Sebastián Elcano (or del Cano, 1486 or 87–1526), returned to Spain, completing the circumnavigation of the world. Although Magellan is often credited with having been the first navigator to circle the globe, that honor really belongs to Elcano.

LORD KELVIN AND HOW TO TAKE SOUNDINGS FROM A MOVING SHIP

Lord Kelvin (1824–1907) was one of the most distinguished British physicists of his generation. In his later years he towered above his contemporaries and inspired awe—sometimes approaching terror in younger scientists. He is buried in Westminster Abbey, next to Sir Isaac Newton (1642–1727).

As well as being a brilliant theorist, Kelvin was also an enthusiast for applying physics to practical problems. His interests were wide, and he was a prolific inventor. During the 1850s several attempts were made to lay a transatlantic telegraph cable (see "Mid-Atlantic Ridge" on pages 58–62). The first successful cable, laid in 1866, was based on Kelvin's suggestion, derived from his extensive knowledge of electrical theory, that fast signals could be transmitted only by using low voltages and using a very sensitive instrument to receive them, such as the mirror galvanometer, which he had invented. He received a knighthood for this contribution to international telegraphy.

Already interested in the sea, in 1870 Kelvin bought a large sea-going sailing boat, the *Lalla Rookh,* partly to distract him from the grief he was suffering from the recent death of his wife, Margaret. The *Lalla Rookh* had luxurious accommodation and also a well-equipped laboratory. Kelvin spent much of his time on the boat conducting research—in 1873 he sailed it to Madeira in connection with his work on laying submarine telegraph cables—and his time at sea stimulated his interest in problems of navigation. He invented and patented a new type of ship's compass, which he tried to persuade the navy to use. Officials at the Admiralty were doubtful about the compass, and one committee thought it would be too flimsy to survive on a warship. The story goes that Kelvin persuaded the committee by throwing his compass across the room. It survived, but when he threw a standard-issue naval compass across the same room, it did not. By 1888 Kelvin's compass was in use on all British naval vessels.

Kelvin also invented a tide gauge, and he solved the problem of taking a depth sounding from a moving ship. In the days of sail this problem was not too serious, because sailing ships were fairly slow, but steamships were much faster, and this made traditional sounding devices inaccurate. When a member of Ferdinand Magellan's crew dropped a sounding line over the side, the slow speed of the ship meant that the weight at the end of the line probably descended almost vertically. It failed to reach the seabed, but if it had done so the length of line paid out would have been a reasonably accurate measure of the depth of water. In the case of a steamship, however, the technique was much less reliable, because as the ship moved forward the sounding line would trail behind it. Consequently, the line descended at an angle from the ship and gave a reading that showed the water to be deeper than it was. In the illustration, the lines descending from the galleon (top) and the steamship (bottom) represent sounding lines to the seabed. As the diagram shows, the difference could be considerable. In order to obtain a true reading, the captain of a steamship had to slow the ship or bring it to a halt, and after taking the sounding the ship had to accelerate back to its cruising speed. This operation took time, delaying the voyage, and captains often dispensed with it and relied on guesswork to estimate the depth of water. The consequence of that was a sharp rise in the number of shipwrecks.

This was the problem Kelvin was asked to address. His solution was a sounding machine that remained in use until the echo sounder replaced it (see "Reginald Fessenden and His Echo Sounder" on pages 54–58). Kelvin entered into partnership with James White (1824–84), owner of a Glasgow company that made optical instruments for the university, to form Kelvin and James White Ltd, which manufactured and marketed what was then called the Kelvin White sounding machine. The company opened a London branch in about 1904, and by 1915 it was trading as Kelvin, White and Hutton Ltd.

The Kelvin White sounding machine used piano wire wound around a small hand-operated winch. The wire was as strong as the hemp rope it replaced, but it could be reeled out and in much faster and, unlike rope, immersion in seawater did not make it sodden and heavy. Simple though it may seem, this was a major innovation—but there was another. The weight, or lead, at the end of a conventional sounding line was a solid cylinder, often hollow at the end and filled with tallow. A lead of this type was said to be "armed." When it reached the bottom, it collected a sample of seabed material. This

Sounding line. The line from the slow-moving galleon is much shorter than the line from the faster steamship.

showed the type of seabed beneath the ship, and it also proved that the lead had reached the bottom. Attached to Kelvin's lead there was a glass cylinder coated on the inside with silver chromate, which is red in color. The cylinder was closed at the top but open at the bottom, so when it entered the water it was filled with air. As the lead sank, increasing pressure forced water into the cylinder, compressing the air. Seawater reacted with the silver chromate to form silver chloride, which is white. When the line was reeled in and the cylinder was recovered—filled with air once more as it left the water—the change in color marked the distance water had penetrated into the cylinder and from that it was a simple matter to calculate the depth. If water had penetrated halfway, the pressure would have been twice the surface atmospheric pressure of about 15 pounds per square inch (101 kPa), which corresponded to a depth of 33 feet (10 m). If water penetrated two-thirds of the way into the cylinder, the pressure was three times atmospheric pressure, and the depth was 66 feet (20 m). Provided the lead reached the bottom, the Kelvin White sounding machine measured the water depth accurately regardless of the length of wire that had been paid out. Steamship captains could take a depth sounding without delaying the ship's progress.

Lord Kelvin had been born in Belfast on June 26, 1824, as William Thomson. His father, James Thomson, was professor of mathematics and engineering at the Royal Belfast Academical Institution. His mother died when William was six years old. Both William and his elder brother James (1822–92) were educated at home by their father. In 1832 William's father was appointed professor of mathematics at the University of Glasgow, and the family moved to Scotland. Two years later, aged 10, William entered Glasgow University to study natural philosophy (as science was then known). It was not unusual for talented students to enter Scottish universities at such a young age because the universities competed with schools for the most able students. Thomson began his degree studies in 1838, when he was 14. In 1841 he transferred to Cambridge University, graduating in 1845. He then spent a short time in Paris before becoming professor of natural history at Glasgow University in 1846. He remained in this post until he retired in 1899.

In the course of his long career William Thomson published more than 600 scientific papers. He was elected to the Royal Society in 1851 and received its Royal Medal in 1856 and its Copley Medal in

1883. He was president of the Royal Society from 1890 until 1895. He was president of the Royal Society of Edinburgh from 1873 to 1878, 1886 to 1890, and from 1895 until his death. Thomson was knighted in 1866, and in 1893 he was raised to the peerage, taking the title of Baron Kelvin of Largs. The Kelvin is the river that runs close to Glasgow University. Largs is the coastal town where he built a fine house. That is where he died on December 17, 1907.

ANTOINE-JÉRÔME BALARD AND THE CHEMISTRY OF THE OCEANS

Over the centuries navigators have developed instruments and techniques to help them cross the oceans safely and arrive predictably at their destinations. Surveyors and cartographers have contributed to this endeavor by charting coastlines and recording the locations of hazards such as hidden rocks and reefs. Geologists have explored the ocean floor. None of these scientists showed any interest in the composition of the oceans, however. They were geographers, physicists, and geophysicists. Not one of them was a chemist.

Chemical analysis involves isolating substances and weighing them precisely. It is a branch of science that did not emerge until the second half of the 18th century, due partly to the work of the Swedish chemist, mineralogist, and naturalist Torbern Olof Bergman (1735–84) and the French chemist Antoine-Laurent Lavoisier (1743–94). Lavoisier placed strong emphasis on the importance of careful observation and accurate measurement. Bergman explored the ways different chemical substances combine and in 1775 published *A Dissertation on Elective Attractions,* a paper in which he included a long table of chemical affinities, arranging elements and compounds in a way that showed which substances would combine with which. Bergman specialized in the chemistry of metals and devised a system for classifying minerals according to their chemical composition, but he also had an interest in seawater. It was Lavoisier who made the first quantitative analysis of seawater in 1772. Bergman analyzed seawater in 1774 but identified only common salt (sodium chloride), magnesium chloride, and calcium sulfate.

Almost a century passed before a French chemist, Antoine-Jérôme Balard (1802–76), began analyzing seawater in search of substances with industrial potential. His first search was for iodine, which had

been discovered in 1811 by another French chemist, Bernard Courtois (1777–1838). Courtois was the son of a saltpeter manufacturer
in Dijon, in eastern France. He helped his father in the factory and
served an apprenticeship as a pharmacist. He studied at the École
Polytechnique and spent some time in the army as a pharmacist, during which he was the first chemist to isolate morphine from opium.
After leaving the army he returned to work in the saltpeter factory.
Saltpeter (potassium nitrate) was a basic ingredient in the manufacture of gunpowder, which was much in demand, for the Napoleonic
Wars were at their height and the blockade of French ports had cut
off the supply of cheap saltpeter from India.

At the Courtois factory the production process began by burning
seaweed, gathered on the coasts of Normandy and Brittany, and boiling the ash in water. This produced a *lye,* which is a strongly alkaline
solution of potassium carbonate. Further processing converted this to
potassium nitrate, but the lye contained many impurities, especially
sulfur compounds. These were removed by adding sulfuric acid and
heating the liquid. One day Courtois inadvertently added too much
acid, and when he heated the liquid it gave off a violet vapor that
condensed on cold surfaces into dark crystals. Courtois examined
the crystals but was unable to investigate it fully and sent samples to
the chemist Joseph-Louis Gay-Lussac (1778–1850) and the physicist
André-Marie Ampère (1775–1836). Ampère passed his specimen to
the English chemist Sir Humphry Davy (1778–1829). Both Gay-Lussac and Davy identified the substance as a new element. Gay-Lussac
suggested calling it *iode,* but it was Davy who gave it the name it has
had ever since. He called it iodine, a word he derived from the Greek
ioeidēs, meaning "violet colored."

Eventually the war ended, the import of Indian saltpeter resumed,
and the Courtois saltpeter business failed. Bernard Courtois abandoned it and struggled to make a living producing and selling iodine.
In 1831 he received a prize of 3,000 francs for his discovery of iodine,
but it was too little, too late. In 1838 he died in poverty in Paris.

Balard was also interested in iodine. He sought other sources of
the element, but at the same time he continued investigating seaweed
and plants from salt marshes and obtaining iodine the way Courtois
had done. When boiling the ash to produce lye, Balard observed that
sometimes the solution turned brown. Curious, in 1826 he isolated
the brown colorant and identified it as a substance with chemical

properties intermediate between those of chlorine and iodine. At first he thought he had a compound of the two, but further investigation revealed it as an element in its own right. The same element had also been discovered independently a year earlier by the German chemist Carl Jacob Löwig (1803–90) using an entirely different method, but Balard was the first to publish his work and is usually credited with the discovery. The German chemist Justus von Liebig (1803–73) had also isolated bromine some years earlier but identified it as iodine chloride. When he learned of Balard's work, Liebig confirmed that his bottle of "iodine chloride" was, in fact, bromine. Bromine vapor has a very distinctive (and unpleasant) odor, and Gay-Lussac suggested the name bromine for it, derived from the Greek *brómos,* which means "stench" (specifically of male goats). Balard continued to be interested in the chemistry of seawater and isolated caustic soda (sodium hydroxide), sodium sulfate, and potash (potassium carbonate) from it. In 1834 he discovered dichlorine oxide (Cl_2O) and chloric acid (HClO). Balard was the first scientist to suggest that bleaching powder was a double compound of calcium chloride and calcium hypochlorite.

AVERAGE COMPOSITION OF SEAWATER	
INGREDIENT	PERCENTAGE BY WEIGHT
Major constituents	
Chloride (Cl^-)	55.04
Sodium (Na^+)	30.61
Sulfate (SO_4^{2-})	7.68
Magnesium (Mg^{2+})	3.69
Calcium (Ca^{2+})	1.16
Potassium (K^+)	1.10
Total major constituents	**99.28**
Minor constituents	
Bicarbonate (HCO_3^-)	0.41
Bromide (Br^-)	0.19
Boric acid (H_3BO_3)	0.07
Strontium (Sr^{2+})	0.04
Total minor constituents	0.71
Total all constituents	**99.99**

Balard was born on September 30, 1802, in Montpellier, in southern France. His parents were poor, and he worked as a laboratory technician at the local faculty of sciences to help pay for his studies at the École de Pharmacie in Montpellier. He qualified as a pharmacist in 1826. It was while still a student that Balard discovered bromine. After graduating, Balard remained at the school of pharmacy as a demonstrator in chemistry. For a time he was a professor there and also at the faculty of sciences. His discovery of bromine brought him to the attention of the leading chemists of the day, and he moved to Paris. In 1842 he was appointed professor of chemistry at the Sorbonne, and in 1851 he became professor of chemistry at the Collège de France. He died in Paris on March 30, 1876.

Today the chemical composition of seawater is well known. It varies somewhat from place to place, but the "Average Composition of Seawater" table lists the major and minor constituents of water with the average salinity for seawater of 34.8‰ (parts per thousand).

BENJAMIN FRANKLIN AND THE "RIVER IN THE OCEAN"

Benjamin Franklin (1706–90) was driven by curiosity. He is known today mainly for his studies of electricity and as a public servant, diplomat, author, and inventor. His scientific interests were not confined to electrical phenomena, however. Everything interested him—including the oceans.

European explorers and settlers had to cross the North Atlantic Ocean to reach North America, and Americans had to cross it to maintain contacts and trading links with Europe. People had been sailing across the ocean in both directions for centuries, and navigators had established the speediest routes to follow. It seems odd, therefore, that it was not until 1769 that any scientist saw fit to investigate the major system of ocean currents that affect journey times.

In 1769 Franklin was in London seeking to improve relations between Britain and its American colonies when he heard there were complaints from the board of customs in Boston to the lords of the Treasury in London about the postal service between Britain and North America. This was of direct concern to him, because since 1753 Franklin had been co-deputy postmaster general for the British North American colonies. The complaint was that mail often took two weeks longer to travel from Falmouth in Cornwall to New York than did merchant ships traveling from London to Rhode Island. The Boston board of customs proposed that henceforth mail packets should sail between London and Rhode Island. This intrigued Franklin because New York and Rhode Island were barely one day's sailing time apart, and while vessels sailing from Falmouth, on the south coast of England not far from Land's End (the westernmost tip of the country), could head immediately into the Atlantic, those sailing from London had to sail all the way down the Thames, around Kent, and the whole length of the English Channel before reaching the open ocean. What is more, merchant ships were larger, heavier, and slower than the mail packets, which were designed for speed. The report made no sense, and at first Franklin thought it must be mistaken. Nevertheless, he decided to investigate.

It so happened that Timothy Folger, a sea captain from Nantucket and relative of Franklin's, was in London at that time, and Franklin consulted him. Folger told him that the report might well be true. The reason, said Folger, was that captains from Rhode Island were very familiar with the Gulf Stream and the captains of packets from England were not. In "Sundry Maritime Observations," a very long letter he wrote in 1785 to Alphonsus le Roy, a French scientific colleague, Franklin recalled what Folger had told him as follows:

> We are well acquainted with that stream, says he, because in our pursuit of whales, which keep near the sides of it, but are not to be met with in it, we run down along the sides, and frequently cross it to change our side: and in crossing it have sometimes met and spoke with those packets, who were in the middle of it, and stemming it. We have informed them that stemming a current, that was against them to the value of three miles an hour; and advised them to cross it and get out of it; but they were too wise to be counselled by simple American fishermen.

Folger went on to say that when winds were light, a ship in the stream would be carried back by the current more than forward by the wind, and even with a favorable wind the stream could slow a ship by 70 miles (113 km) a day. Franklin remarked that it was strange the current was not marked on nautical charts. He asked Folger to mark it on a chart, which he did, adding directions for avoiding it when sailing from Europe to North America. Franklin arranged for the chart to be engraved and printed in London and for copies to be sent to the captains of the mail packets in Falmouth. They ignored it, however.

Although Juan Ponce de Léon (1460–1521), the Spanish explorer and discoverer of Florida, had known of the existence of the Gulf Stream (Franklin called it the Gulph Stream), Franklin and Folger were the first people ever to chart it. Franklin's interest did not end with marking the position of the Gulf Stream, however. He speculated about what caused it as follows:

> This stream is probably generated by the great accumulation of water on the eastern coast of America between the tropics, by the trade winds which constantly blow there. It is known that a large piece of water ten miles broad and generally of only three feet deep, has by a strong wind had its waters driven to one side and sustained so as to become six feet deep, while the windward side was laid dry. This may give some idea of the quantity heaped up on the American coast, and the reason of its running down in a strong current through the islands into the Bay of Mexico, and from thence issuing through the gulph of Florida, and proceeding along the coast to the banks of Newfoundland, where it turns off towards and runs down through the western islands.

Franklin had surmised, correctly, that the current is driven by the wind and that the height of sea level is not everywhere the same, so the current flows down a slope. Oceanographers now know that the sea level around Bermuda, where the current has moved away from the American coast, is 3.3 feet (1 m) higher than it is along the eastern coast of the United States. Franklin also routinely measured the sea temperature during his many crossings of the Atlantic, and he found that the water of the Gulf Stream is warmer than the water outside it; in fact, the difference off the eastern coast of the United States averages about 18°F (10°C). The warm water heats the air above it, causing it to rise, and this, Franklin suggested, was the cause of "those tornados and waterspouts frequently met with, and seen near and over the stream," and where it meets cold air in the vicinity of Newfoundland that it was also responsible for the fog that is frequently found there. In addition, he examined the seaweeds growing in the warm water.

It was Benjamin Franklin who first called it the Gulph (Gulf) Stream. In *The Physical Geography of the Sea* published in 1855, Matthew Maury (see "Matthew Fontaine Maury, Ocean Currents, and International Cooperation" on pages 29–33) described it more romantically. "There is a river in the ocean," he wrote. "In the severest droughts it never fails, and in the mightiest floods it never overflows; its banks and its bottom are of cold water, while its current is of warm; the Gulf of Mexico is its fountain, and its mouth is the Arctic Sea. It is the Gulf Stream. There is in the world no other such majestic flow of waters."

Benjamin Franklin was born on January 17, 1706, in Boston, Massachusetts, the 15th of the 17 children of a soap and candle maker who had emigrated from Banbury, Oxfordshire, in England. The family could not afford to pay for his education, and Franklin spent only one year at a grammar school. He received some private tuition but was mainly self-taught. In 1717 Benjamin began an apprenticeship as a printer under his elder brother James. James founded a newspaper, *New-England Courant,* in 1721. Benjamin wrote for the paper and took over as publisher when James's criticism of the authorities finally landed him in jail. After quarreling with James, Franklin moved to New York, found no work there, and moved on to Philadelphia, where he was more successful, and later to London. He became a printer and publisher with a business from which he made enough money to retire in 1748. He was then able to devote more of his time to his scientific experiments. In 1753 he received the Copley Prize from the Royal Society and honorary master's degrees from Harvard and Yale Universities in 1753 and the College of William and Mary in 1756, a doctorate in law from the University of Edinburgh in 1759, and a doctorate in civil law from the University of Oxford in 1762.

From 1753, when he became an official of the post office, Franklin's life began to center on public service. He was an agent for Pennsylvania, Georgia, New Jersey, and Massachusetts, and in 1776 he went to Paris as part of a commission hoping to secure military and economic assistance in the struggle for American independence. Franklin had become a diplomat, and before long he was a hero throughout France, responding so well to French admiration that his popularity lasted for several generations. He returned home in

1785, starting his journey by being carried to the French port of Le Havre in one of Queen Marie Antoinette's own litters. It was during that voyage that he wrote "Sundry Maritime Observations." Back in Philadelphia he was elected to the executive council of the state and a few days later became its president, serving for three years. His health was failing, however, and by 1789 he was bedridden and reliant on opium to control his pain. His last public act was to support a campaign to have the first Congress of the United States consider the abolition of slavery. Franklin died on April 17, 1790. His funeral in Philadelphia was the most splendid the city had ever seen, and in France there were many eulogies to the man the French regarded as a symbol of enlightenment and freedom.

JAMES RENNELL, WHO MAPPED THE ATLANTIC CURRENTS

Upcott is a village not far from the market town of Chudleigh, in Devon, England. The Royal Geographical Society in London holds in its archives a map of Upcott. It is one of the first items the society acquired after it was founded in 1830, shortly after the death of Major James Rennell (1742–1830), who had drawn the map while still at school. Rennell was Britain's leading geographer and also, according to Sir Clements Markham (1830–1916)—a later president of the Royal Geographical Society—in his biography of Rennell, "the founder of oceanography: that branch of geographical science which deals with the ocean, its winds and currents." Rennell charted the systems of ocean currents in the Atlantic and Indian Oceans. In his last published work, *Currents of the Atlantic Ocean,* published posthumously in 1832 by his daughter Jane, Rennell described the Gulf Stream as an immense river descending from a higher level into a plain. This was the metaphor Maury used and expanded in his *Physical Geography of the Sea.*

Rennell began charting the currents in 1810, when he was 68 years old, and he devoted the last 20 years of his life to the task. No one could have attempted the task earlier because it was not until then that navigators had the chronometers they needed to calculate longitude accurately. Until the late 18th century the most accurate chronometers were regulated by the motion of a pendulum swinging under the force of gravity, but this meant the instruments had to stand on a stable, level surface. The rolling of a ship at sea made them useless. In 1761 the Yorkshire carpenter John Harrison (1693–1776) constructed the first marine chronometer that was robust and accurate enough for the task, but it was not until early in the 19th century that naval vessels began to be routinely supplied with them.

In order to chart the course of a current, at regular intervals a ship's navigator must compare the position of the ship calculated by dead reckoning with its actual position. Dead reckoning is the technique of calculating a position from the compass heading, speed, and elapsed time since the last known position. It tells the navigator where the ship will be, provided its progress is governed by only these factors. A current will accelerate or slow the ship, and unless it flows in the same or opposite direction to the ship, it will deflect the ship from its course. If the true position of the ship is known, the strength and direction of the current can be calculated from the difference between the true and dead-reckoning positions.

Measuring the angle of the Sun above the horizon at noon will reveal latitude. On every day of the year the Sun is directly overhead at some latitude between the tropics of Cancer and Capricorn, so the latitude of the ship (L) is equal to $90 - A \pm D$, where 90 is the latitude of the North and South Poles, A is the angle of the Sun above the horizon, and D is the correction to take account of the position of the Sun on that date, adding or subtracting depending on which hemisphere the ship is in. So, given the measured height of the Sun above the horizon, the date, and knowledge of whether the ship is in the Northern or Southern Hemisphere, the navigator can calculate the latitude or look it up in a book of tables.

Measuring longitude is where the chronometer becomes essential. The Earth rotates on its axis through 360° every 24 hours, which means it turns through 15° (360° ÷ 24) every hour. If the navigator notes the precise moment when the Sun reaches its highest point in the sky (noon) and compares that time with the time shown by a chronometer set to show the time at the 0° meridian, the difference between the local and 0° time measured in hours and multiplied by 15 (and subtracting 180 if the result is larger than 180) gives the longitude. The accuracy of the calculation depends on that of the chronometer set to 0° time.

Rennell charted the meanderings of the Gulf Stream, with its huge eddies swirling around cold water at their centers. He based his studies on data gathered from the notebooks and logbooks supplied to him by his many friends. He also charted the Agulhas Current, which flows southward off the eastern coast of southern Africa, and explained the cause of the northward current, now known as Rennell's Current, that intermittently flows to the south of the Isles of Scilly, off the southwestern tip of Britain.

James Rennell was born on December 3, 1742, at Upcott. His father, an army officer, was killed in action soon after James was born, and James was raised by a guardian, the Reverend Gilbert Burrington. In 1756, at the age of 14, Rennell joined the navy as a midshipman. Britain was then fighting in the Seven Years' War (1756–63), and Rennell saw action in several sea battles before being sent to India in 1760. There he learned marine surveying, and his captain lent him to the East India Company as a surveyor. He served on one of their ships for a year, sailing to the Philippines and drawing maps of several of the harbors they visited along the way. When the Seven Years' War ended Rennell saw no prospect of promotion in the Royal Navy, so he joined the sea service of the East India Company and was immediately given command of a ship. Unfortunately, almost at once his new command was destroyed by a tropical cyclone. Rennell was ashore at the time and so was unharmed, and he was given command of a small yacht, the *Neptune,* in which he surveyed the coastline of southeastern India. He was then sent to Calcutta (now Kolkata), where he was appointed surveyor-general of the East India Company territories in Bengal, with a commission in the Bengal Engineers. His was the first-ever survey of Bengal. In 1776 Rennell was seriously wounded in a skirmish on the border of Bhutan. He never recovered fully and retired in 1777 with the rank of major. He returned to London and devoted the rest of his life to geographical and oceanographical research.

Rennell was elected a fellow of the Royal Society in 1781 and received the society's Copley Medal in 1791. The Royal Society of Literature presented him with its gold medal in 1825. In 1772, while still in India, Rennell had married Jane Thackeray. They had two sons, Thomas and William, who died in 1846 and 1819, respectively, and a daughter Jane, who married Admiral Sir John Tremayne Rodd and who died in 1863. James Rennell died on March 29, 1830, and was buried in the nave of Westminster Abbey, where there is a tablet to his memory.

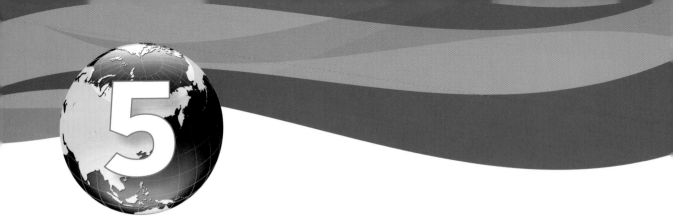

Journeys to the Bottom of the Sea

When a valuable object falls into the sea or a vessel sinks taking its cargo with it to the seafloor, the only way to recover the lost items is for a human to dive for them. There are other seabed resources that coastal communities have been exploiting for thousands of years. Shellfish such as oysters, clams, and scallops live below the surface. Traditionally, divers swam down to gather them, and the pearls oysters sometimes contain were a profitable bonus. Sponges were also valuable, and they, too, live on the seabed.

Obviously, it is impossible for a human to breathe underwater, so without an air supply a dive can last for only as long as the diver is able to hold his or her breath. Most people can hold their breath for about 40 seconds, but since ancient times in lands around the Mediterranean and in Asia, professionals diving from ships for pearls and sponges have trained themselves to hold their breath for up to several minutes. They overcame the natural buoyancy of the human body by holding heavy stones that they released when they were ready to ascend, and a rope around their waists attached them to the ship to guide their ascent and haul them to the surface in an emergency.

Divers were used in ancient Greece to retrieve sunken treasure and similar items of commercial or military importance. The earliest written mention of diving is in book 16 of the *Iliad*, Homer's epic story of the Trojan War written in about 750 B.C.E. This passage tells how the warrior Patroclus killed Cebriones by throwing a sharp stone that struck him on the forehead, hurling him from his chariot.

As Cebriones fell, Patroclus jeered: "Ha! Quite an acrobat, I see, judging by that graceful dive! The man who takes so neat a header from a chariot on land could dive for oysters from a ship at sea in any weather and fetch up plenty for a feast. I did not know that the Trojans had such divers." Divers were also used to attack enemy ships and heavily defended coastal cities.

Lack of air is only one of the problems that divers face. The others are lack of visibility and pressure. Seawater absorbs and scatters light, so visibility deteriorates with increasing depth. The pressure on a diver's body increases rapidly with distance from the surface, making it increasingly difficult to prevent air being forced out of the lungs.

This chapter tells of the invention and development of equipment and machines that have allowed divers to work on the seabed and of submersibles in which scientists have descended into the deepest ocean trenches. It also explains the physiological effects of extreme pressure and how divers survive them.

DIVING BELLS AND DIVING SUITS

The most obvious solution to the lack of air below the sea surface is to equip the diver with a breathing tube, like a longer version of the snorkel modern swimmers use. The diver holds one end of the open tube in his or her mouth, and the other end is held above the surface, perhaps by a float. Many people thought of this, but found it does not work because pressure rises rapidly as a diver descends (see the sidebar "Pressure and Depth in the Ocean"). Snorkelers swim at the surface, but in order to breathe below the surface the diver must exert a force to expand the lungs that is greater than the pressure exerted by the water. At a depth of about seven feet (2 m) this is extremely tiring. Below that depth it is impossible, because human chest muscles are simply not strong enough.

There was an alternative, however, and ancient Roman divers used it: a bell-shaped vessel, weighted around the rim, that was lowered vertically into the water. The bell was full of air, and divers could swim into it to breathe. These were the first diving bells, and they remained in use until the 19th century. In 1690 the English astronomer Edmond Halley (1656–1742) invented a method for recharging the air inside the bell from a barrel of air lowered from the surface. Some years later the English inventor John Smeaton (1724–92) improved on this by linking the bell to the ship by a tube and installing a pump on the ship to keep fresh air flowing into the bell.

Diving bells were heavy and cumbersome, however, and they restricted the movements of the divers, who could not move far from their source of air. During the 18th century various inventors tried to overcome this difficulty by having the diver carry the bell as part of a diving suit. In about 1715 Pierre de Rémy, chevalier de Beauve, a

PRESSURE AND DEPTH IN THE OCEAN

Air and water are fluids and possess weight. The weight of the air that forms the atmosphere exerts a pressure at the Earth's surface. This is measured as the pressure exerted on a unit area by the weight of a column of air extending all the way to the top of the atmosphere. The density of air varies according to temperature, because air expands, becoming less dense, when its temperature increases, and contracts, becoming denser, when it cools. At sea level when the air temperature is 59°F (15°C), the average atmospheric pressure is 14.7 pounds per square inch (101.325 kilopascals [kPa]). That is the average weight of a column of air with a cross-sectional area of one square inch (or 1 m², if measured in kPa).

Water also exerts pressure. This increases with depth, because as with air, the pressure represents the weight of overlying water acting on a unit area of surface. Water is much denser than air: One cubic foot of seawater weighs 64 pounds (1 m³ weighs 1.027 tonnes [t]). Consequently, water pressure increases rapidly with increasing depth. The water pressure at any given depth is calculated using the *hydrostatic equation*. This is $p = g\rho z$, where p is pressure; g is acceleration due to gravity (32.175 feet per second per second [9.807 m/s²]); ρ is the density of water (for seawater 64

pounds per cubic foot [1.027 t/m³]); and z is the depth below the surface. The hydrostatic equation gives the pressure exerted by the water. Unless the water surface is covered, sealing the water from the atmosphere, the total pressure below the surface is equal to the water pressure plus the atmospheric pressure acting on the water surface.

The result of the calculation shows that seawater pressure increases by approximately 14.7 pounds per square inch for every 33 feet of depth (101.325 kPa/10 m). Adding the atmospheric pressure to the water pressure, the following table shows the pressure at various depths in the ocean.

PRESSURE CHANGE WITH DEPTH

DEPTH		PRESSURE	
Feet	Meters	Pounds/inch²	kPa
0 (surface)	0	14.7	101.325
33	10	29.4	202.650
66	20	44.1	303.975
99	30	58.8	405.300
132	40	73.5	506.625
165	50	88.2	607.95

guardsman in the French navy stationed at Brest, Brittany, invented one of the first diving suits. The diver wore sandals with lead soles and a waterproof suit with a helmet attached to rigid breast and back plates that resisted water pressure. There were two tubes to supply and remove air. In 1772 the French inventor Sieur Freminet made a diving suit in which he descended to 50 feet (15 m) at Le Havre. The suit was made of leather, and the helmet was of copper. The diver carried compressed air in a container on his back, connected to the helmet by two tubes. Freminet's description claims that the suit "regenerated" exhaled air. He made several dives, each time remaining underwater for several minutes. His device merely retained his exhaled air, however, and he died from lack of oxygen after remaining submerged for 20 minutes.

Inventors hope to make money from their contraptions, and the aim of all these diving bells and suits was to allow men to perform useful work on the seabed. As if to emphasize this, the picture of the suit invented in 1797 by the German engineer Karl Heinrich Klingert (1760–1828) shows the diver holding an axe, perhaps to break into the wreck of a wooden ship in search of its valuable contents. Klingert's suit had two tubes connected to an air tank. The air tank contained an adjustable piston that made it possible to control buoyancy, and it was suspended in the water behind the diver. Although the suit worked and several successful test dives were made with it, it was so cumbersome that Klingert later modified it, making the air tank smaller.

In 1819 the German inventor August Siebe (1788–1872) designed the diving suit that with later modifications remained in use for more than 150 years. Siebe suits were still being manufactured in 1975. Its most famous feature was the "Twelve Bolt Helmet," named for the number of bolts that

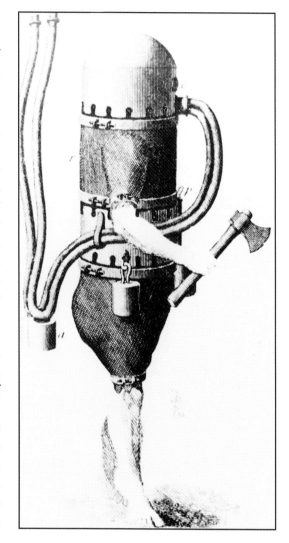

The diving suit invented in 1797 by the German engineer Karl Heinrich Klingert (1760–1828). The domed cylinder, made from tinplate, covered the diver's head and torso, leaving his arms free, and tight straps at the elbows and knees kept the suit watertight. Weights hung at the waist provided ballast, and the diver could see out through a glass window. There are two tubes, one feeding fresh air under pressure and the other evacuating exhaled air. *(Science Photo Library)*

projected from the metal shoulder piece, called the corselet. The diver wore a waterproof suit made of canvas and rubber that was secured to the corselet by the 12 bolts, making a waterproof seal. The spherical helmet, with round windows in front and on either side, was then screwed into place on the corselet. Air was supplied from a pump at the surface, and a line attached the diver to the ship. The diver wore heavy boots and lead weights attached around the waist.

In the first version, the suit filled with water, the air pressure inside the helmet ensuring that the water did not rise as far as the wearer's head. In 1836 Siebe changed the design to one with a suit that kept the wearer dry. This was important, because seawater is cold.

At first the diver communicated with the surface by tugging on the line. Then speaking tubes were added, through which the diver and the person operating the line and air pump could shout to each other. Eventually a telephone line replaced the speaking tube. Heavy and cumbersome though it looked, the resulting suit was light and easy to wear underwater, and divers could be trained quickly. The navies of many countries adopted the suit, and it was also widely used by civilian companies.

PAUL BERT AND THE BENDS

Divers who swim down in search of shellfish, sponges, or pearls remain underwater for only a few minutes, dive only in shallow water, and they return to the surface unharmed. Diving suits allowed divers to descend to greater depths and to remain submerged longer. That is when a new problem emerged. Sometimes divers returning to the surface suffered from pain deep inside their joints. They might be confused, have trouble with their vision, feel nauseated, complain of chest pains, and have difficulty breathing. The severity of the symptoms varied, but in extreme cases they could be fatal. Because they could be doubled up with pain, the divers called this condition the *bends*; it was also called *caisson disease*, and nowadays its medical name is *decompression sickness*. (Caissons are pressurized chambers or tunnels in which workers have access to construction areas in tunnels and at the supports of bridges.)

The sickness has been known for a long time, and it is not confined to divers: Miners working deep below ground can also suffer from it, as do aviators who climb too rapidly. It is caused by a rapid reduction in pressure, and as long ago as 1670 the Irish physicist Robert Boyle (1627–91) demonstrated that this could cause bubbles to form in animal tissue. It was not until the late 19th century that a scientist discovered why the bubbles formed and how they caused pain.

That scientist was the French doctor Paul Bert (1833–86), professor of physiology at the Sorbonne (University of Paris), in Paris. In 1875 the Academy of Sciences awarded Bert a bounty of 20,000 francs to find the cause of the bends. He published his results in 1878 in a book entitled *La pression barométrique: Recherches de physiologie expérimentale* (published in English in 1943 with the title *Barometric Pressure: Researches in Experimental Physiology*).

When a person breathes, oxygen in the air diffuses across cell membranes in the lungs and dissolves in the blood. Oxygen accounts for only about 21 percent of air by volume, however. Seventy-eight percent is nitrogen. Bert had discovered that as pressure inside the lungs increases, nitrogen also begins to cross the cell membranes and is forced into solution. Blood carrying dissolved nitrogen then travels to every part of the body, including the joints. If the high pressure is reduced slowly, the nitrogen gently diffuses out of solution and is exhaled. If the pressure is reduced too rapidly, however, the nitrogen comes out of solution so fast that it forms bubbles. The bubbles lodge in the fine capillary blood vessels, and it is the pressure they exert that causes the symptoms of the bends. The process is similar to the way fizzy drinks are made: Carbon dioxide is forced into solution under high pressure, and when the can or bottle is opened, reducing the pressure, the gas bubbles out of solution.

Bert advised that divers should return to the surface slowly enough to allow the nitrogen to leave their blood harmlessly. If they start to experience the symptoms of the bends, the remedy is to return them to the pressure they were under at depth, which forces the nitrogen back into the blood, and then reduce the pressure slowly. He also found experimentally that breathing pure oxygen helped relieve the symptoms of the bends and suggested, correctly, that nitrogen diffuses rapidly and smoothly into pure oxygen.

Between 1892 and 1900 the Scottish physiologist John Scott Haldane (1860–1936) developed methods for studying respiration and the blood, and in 1905 the Deep Diving Committee of the Royal Navy asked him to investigate the bends. Haldane found that halving the pressure on the diver would not produce symptoms. It meant that a diver 130 feet (40 m) below the surface could rise rapidly to half that depth, 65 feet (20 m), without experiencing the bends. If the diver remained at this depth long enough for excess nitrogen to leave the blood, a rise to half that depth, 33 feet (10 m), would similarly produce no ill effects. Bert had advocated uniform decompression, achieved by rising continuously at a constant speed. This was simple, but very slow, and it was easy to miscalculate the rate of ascent. Haldane recommended decompression in stages, which was much quicker and also safer. His 15-year-old son, J. B. S. Haldane (1892–1964), compiled decompression tables that allowed those bringing divers to the surface to time the stages of their ascents correctly.

The elder Haldane had discovered that different body tissues absorb nitrogen at different rates. He divided tissues into five basic types, which he called compartments, and for each one he measured the time it took to absorb and eliminate an amount of nitrogen that reduced the nitrogen pressure by half (the half-life). These times varied from five to 75 minutes. Modern decompression tables are based on calculations from six to 12 compartments, with half-lives of five to 240 minutes.

Paul Bert was born on October 17, 1833, at Auxerre, in the Burgundy region of France to the southeast of Paris. He enrolled at the École Polytechnique in Paris with the intention of studying engineering, but changed his mind and began studying law. Then he fell under the influence of the zoologist Louis-Pierre Gratiolet (1815–65) and switched again, this time to physiology. He graduated as a doctor of medicine in 1863 and as a doctor of science in 1866. In 1866 Bert was appointed professor of physiology at the University of Bordeaux, and in 1869 he became professor of physiology at the Sorbonne.

As well as being an eminent physiologist, Bert was also politically active—from November 14, 1881, to January 30, 1882, he was minister of education and worship in the French government. In early 1886 he was appointed resident-general in Annam and Tonkin, parts of Vietnam, which was then a French colony. He died from dysentery in Hanoi on November 11, 1886.

John Haldane was born in Edinburgh on May 3, 1860, and educated at Edinburgh Academy and went on to study medicine at the University of Edinburgh, graduating in 1884. He then spent a short time in Germany at the Friedrich Schiller University of Jena and the University of Berlin. He was a demonstrator in physiology at the

University of Dundee and later Oxford, and was a reader at Oxford from 1907 until he resigned in 1913 to become director of the Mining Research Laboratory in Doncaster, Yorkshire. The laboratory moved to Birmingham in 1921, and Haldane remained its director until 1928. He lectured at Yale in 1916, at the University of Glasgow from 1927 to 1929, and at the University of Dublin in 1930.

Some physiologists believe the only way they can understand the effects of certain substances and treatments is to try them on themselves. Haldane was famous for experimenting on his own body. He would lock himself in sealed chambers while inhaling toxic gases and making notes of his sensations. During World War I he visited the front to investigate the poison gases being used there; this resulted in his invention of the gas mask.

Haldane married Louisa Kathleen Trotter. They had two children, John Burdon Sanderson Haldane, a physiologist who became as famous as his father, and the novelist and poet Naomi Mitchison (1897–1999). Haldane died at Oxford on the night of March 14–15, 1936.

CHARLES WILLIAM BEEBE AND THE BATHYSPHERE

In 1906 J. B. S. Haldane (see "Paul Bert and the Bends" on pages 108–111) dove to 200 feet (60.96 m) while collecting data for his decompression tables. Some divers have reached slightly greater depths, but the huge water pressure limits the depth attainable by a diver wearing a suit made from flexible material. A diver wearing an atmospheric diving suit (ADS) can descend deeper. An ADS is rigid and maintains an internal pressure equal to surface atmospheric pressure regardless of the external pressure.

The ADS is an old idea. The earliest was in use in 1715, when it made several dives to 60 feet (18 m). The first suit with articulating joints was made in Britain 1838 by W. H. Taylor, but it would certainly have failed if it had descended very far. Lodner D. Phillips, of Chicago, designed the first completely enclosed armored suit, patented in 1856, although there is no record of it ever being built. Several other designs were proposed during the 19th century, each improving on the last, until Joseph Salim Peress (1896–1978) built JIM 1—he called it the Tritonia—in 1932. Jim Jarrett, chief diver for Peress and the Jim whose name was given to the suit, dove on the wreck of the *Lusitania* in JIM 1, reaching 312 feet (95 m). In May 1930 Jarrett dove to 403 feet (123 m) in Loch Ness, Scotland. At that depth the water pressure is approximately 195 pounds per square inch (1,348 kPa).

JIM suits and other atmospheric diving suits are still in use, and modern versions have thrusters to help the wearer move and to improve control in currents. An ADS places a worker directly on a site, and it can rise and descend as rapidly and as often as necessary, with no need for pauses to allow nitrogen to leave the diver's blood. Its disadvantage is that the diver is unable to carry very much, and its handling tools provide no feedback, so they are clumsy and difficult to use. JIM suits are used mainly in the offshore oil and gas industries and to a lesser extent in salvage.

Meanwhile other inventors were taking a different approach. In 1928 the American inventor, diver, and actor Frederick Otis Barton Jr. (1899–1992) invented the idea of the *bathysphere;* the name is derived from two Greek words, *bathos* (meaning depth), and *sphaira* (meaning sphere). A bathysphere is an unpowered spherical vessel that is lowered into the ocean on a cable from a winch on the deck of its support ship. The actual vessel was designed and built in 1929 by Cox & Stevens, Inc., a yacht design and brokerage company in New York City. It consisted of a hollow sphere 4.75 feet (1.45 m) in diameter made from steel one inch (2.54 cm) thick. The occupants entered through an entrance hatch weighing 400 pounds (176 kg) that was secured by bolts, and once on board they viewed the world through windows made from fused quartz that were three inches (7.62 cm) thick. They breathed oxygen from a pressurized cylinder, and electric fans circulated the air inside the vessel, making it pass over a pan of *soda lime*—a mixture of about 75 percent calcium hydroxide ($Ca[OH]_2$); 20 percent water (H_2O); 3 percent sodium hydroxide ($NaOH$); and 1 percent potassium hydroxide (KOH). Soda lime removed carbon dioxide (CO_2) from the air. A separate pan contained calcium chloride ($CaCl$), which removed moisture from the air. A rubber hose carried telephone wires and the electrical supply from the ship to the bathysphere. Barton's bathysphere is displayed at the New York Aquarium in Coney Island, Brooklyn.

Barton enlisted the help of the American explorer and naturalist William Beebe (1877–1962), and together they made the first descent in the bathysphere on June 6, 1930, reaching a depth of

803 feet (245 m). On August 15, 1934, during their third season of dives, the two men dove to 3,028 feet (923.5 m) off Nonsuch Island, Bermuda, setting a depth record that remained unbroken for 15 years. The photograph shows Beebe with the bathysphere on his return to New York from that dive. As their craft descended beyond the depth light can penetrate, the bathysphere's spotlight revealed clear water, with giant shadows suggesting the presence of huge animals. Beebe described the twinkling lights of *bioluminescent* fish—fish that produce light without heat from colonies of bacteria inhabiting specialized organs (see the sidebar "Bioluminescence" on page 136)—and strange fish he had never before seen alive and swimming.

It was Barton who broke the bathysphere's 1934 record. He had designed a vessel he called the *benthosphere*; the Greek word *benthos* means depths of the sea. The benthosphere resembled the bathysphere and was about the same size, but its steel was 1.75 inches (4.45 cm) thick, allowing it to withstand pressures down to 10,000 feet (3,050 m), and it had two fused quartz windows, one facing horizontally and one facing vertically downward. In August 1949 Barton made

William Beebe (left) and his associate John T. Vann, shown here standing beside the bathysphere on their arrival at New York on November 2, 1934, after descending to a record depth of 3,028 feet (923.5 m) *(AP Images)*

a solo descent to 4,500 feet (1,372.5 m). This remains the greatest depth attained by a submersible suspended from a surface vessel by a cable. Beyond this depth the weight of the cable makes it impossible to control the submersible. Any explorer seeking to descend deeper must do so in a self-propelled vessel.

Charles William ("Will") Beebe was born on July 29, 1877, in Brooklyn, New York City. The family moved several times during his childhood, finally settling in East Orange, New Jersey. Both his parents were keen naturalists, often visiting the American Museum of Natural History and attending lectures there, and Beebe grew up with a love of nature. He attended East Orange High School and Columbia University (1896–99), where Henry Fairfield Osborn (1857–1935), president of the museum, was a professor. There is some doubt whether Beebe graduated from Columbia, but he later received honorary doctorates from Tufts College (now Tufts University) and Colgate University.

In October 1899 Beebe started work as assistant curator of birds at the Zoological Park, which had been recently opened by the New York Zoological Society. He became full curator in 1902, and the same year he married Mary Blair Rice (1880–1959). Will and Mary traveled to Mexico to study birds and obtain specimens for the zoo, describing their experiences in *Two Bird-Lovers in Mexico,* published in 1905. This was the first of Beebe's popular books on natural history and travel. Will and Mary divorced in 1913 (Mary went on to become a novelist), and in 1927 Beebe married Elswyth Thane Ricker (1900–81), an established novelist (who wrote under the name Elswyth Thane). When the Zoological Park opened a department of tropical research, Beebe became its director, and in 1916 he established the zoo's first research station, at Kalacoon, British Guiana (now Guyana).

During the 1920s Beebe became interested in ocean diving, at first using a diving suit. He joined several expeditions by the New York Zoological Society's vessel the *Arcturus,* hauling aboard specimens caught by net in deep water that failed to survive the reduced pressure at the surface. Beebe wished to see these animals alive and in their natural habitat, but it was impossible to reach the depths where they lived wearing a diving suit. It was in 1928, while seeking a practical design for a deep-ocean submersible, that he met Barton.

Beebe retired from the Department of Tropical Research in 1952 and moved to the Arima Valley in Trinidad, where he died on June 4, 1962.

AUGUSTE PICCARD AND THE BATHYSCAPHE

A bathysphere cannot reach the deepest regions of the ocean floor because the weight of its suspension cable renders the vessel uncontrollable. An explorer who wishes to probe deeper waters requires a submersible that can move independently of its support ship. In the 1930s the first such submersible was designed by the Swiss physicist and engineer Auguste Piccard (1884–1962), professor of physics at the Free University of Brussels. Work on its construction began in Belgium in 1937 but had to be halted during World War II. The vessel was finally completed between 1946 and 1948. It was called the *FNRS-2*, for the Belgian Fonds National de la Recherche Scientifique (National Funds for Scientific Research), the Belgian agency that financed it. (The same body also funded *FNRS-1*, which was a high-altitude balloon, also designed by Piccard.) In 1948 the *FNRS-2* was damaged during its trials in the Cape Verde Islands, and the vessel was largely rebuilt and improved. The new version, the *FNRS-3*, was sold to the French Navy. On February 15, 1954, *FNRS-3* descended to 13,000 feet (4,000 m) off Dakar, Senegal.

Experience gained with the *FNRS-2* and *FNRS-3* allowed Piccard to design their improved replacement. The observation gondola for this submersible was made in two sections and built in central Italy. The upper part of the vessel was built in Trieste, a seaport on the shore of the Adriatic Sea in northeastern Italy, and the city gave the vessel its name: *Trieste*. It was launched on August 1, 1953, and later that year dove to more than 10,300 feet (3,150 m). The U.S. Navy bought the *Trieste* in 1958. The photograph shows the USS *Trieste* shortly before its record-breaking dive into the Mariana Trench (see "The *Trieste* and Its Voyage to the Challenger Deep" on pages 119–122).

The bathyscaphe had two sections. The two persons it carried were crowded into the spherical observation gondola, 6.5 feet (2 m) in diameter. The gondola was mounted beneath the much larger float. The float was necessary because in order to withstand the pressure at the depths it was intended to reach, the steel of the gondola was too

Auguste Piccard (1884–1962), wearing a life jacket, is seen climbing from the conning tower of the bathyscaphe *Trieste* in this photograph taken on October 3, 1953. The *Trieste* has just surfaced after diving to 10,335 feet (3,150 m) near Ponza Island off the west coast of Italy. *(Keystone/Stringer)*

thick for the sphere to float unaided. The gondola had one window, facing forward and down. A powerful flashlight was mounted on the outside of the sphere above the window, and there were floodlamps forward of the forward ballast hopper. The gondola carried compressed air that was recycled through a *rebreather* similar to those used in spacecraft, in which exhaled air passes through cylinders of soda lime and calcium chloride, as in the bathysphere (see "Charles William Beebe and the Bathysphere" on pages 111–115).

The float held tanks filled with gasoline to provide buoyancy. Gasoline was chosen because it was easily obtainable, less dense than water, and (like all liquids) almost incompressible. The fact that it can be only slightly compressed allowed it to be contained in relatively thin-walled tanks; as the external pressure increased, the gasoline's incompressibility prevented the tanks from being crushed.

The gasoline tanks were open at the bottom. As the bathyscaphe descended some seawater entered the flotation tanks, maintaining the gasoline at constant pressure. The float made the vessel neutrally buoyant. If necessary, small amounts of gasoline could be released to be replaced by water and increase the vessel's weight.

At either end of the bathyscaphe were air tanks. In order to descend, vents on the tops of these tanks were opened, allowing seawater to enter, thereby increasing the vessel's weight. This is the method a conventional submarine uses to descend, but a submarine differs from a bathyscaphe by filling its buoyancy tanks with compressed air in order to ascend. A bathyscaphe cannot do this, because at the depths it is designed to reach the external water pressure is too great for a pump to be able to expel water. Instead, the bathyscaphe had hopper tanks fore and aft of the gondola, filled with iron pellets. Once below the surface, the hoppers were opened at the bottom, and the iron shot was then held in place by electromagnets at the throats of the hoppers. When the vessel was to rise, the magnets were switched off, releasing the shot. In the event of a power failure, the magnets would fail, the shot would be released automatically, and the bathyscaphe would ascend. A guide rope, in fact made from heavy chain, hung below the bathyscaphe. As the vessel approached the seafloor the chain made contact with the surface, and as more and more of the chain lay on the surface the weight of the bathyscaphe decreased by the weight of the settled chain. This simple mechanism slowed the submersible's descent and prevented it from landing too heavily.

The crew entered the vessel through a structure resembling a submarine conning tower on the upper deck of the float. This gave access to a tunnel leading vertically downward through the float and to a hatch on the gondola. A snorkel tube ran upward from the gondola to above the float, allowing air to be exchanged while the vessel was on the surface. The bathyscaphe was powered by electric motors driving propellers on the upper deck of the float, and a large rudder provided directional control. The illustration shows the general layout.

Auguste Antoine Piccard and his twin brother, Jean-Félix, were born at Basel, Switzerland on January 28, 1884. Their father, Jules Piccard, was a professor of chemistry at the University of Basel. Both boys were educated in Basel and enrolled together at the Swiss Federal Institute of Technology in Zurich, Auguste to study physics and Jean to study chemistry and aeronautical engineering. Both

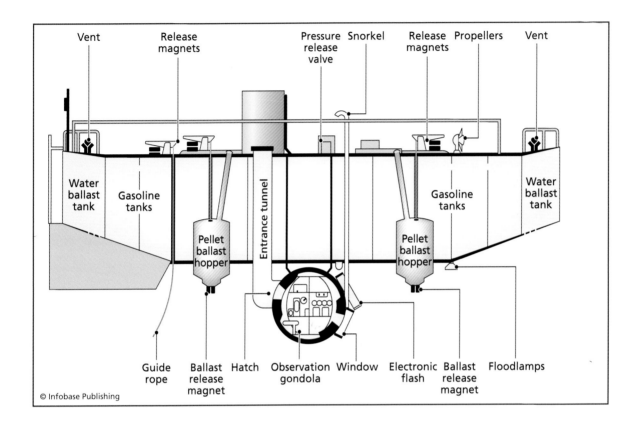

Vent   Release magnets   Pressure release valve   Snorkel   Release magnets   Propellers   Vent

Water ballast tank   Gasoline tanks   Entrance tunnel   Pellet ballast hopper   Pellet ballast hopper   Gasoline tanks   Water ballast tank

Guide rope   Ballast release magnet   Hatch   Observation gondola   Window   Electronic flash   Ballast release magnet   Floodlamps

© Infobase Publishing

This schematic drawing shows the layout of the bathyscaphe *Trieste* at about the time it was acquired by the U.S. Navy.

graduated as doctors of science. Jean became a chemist and designer of balloons. He moved to the United States and became an American citizen in 1931. He died in Minneapolis, Minnesota, on January 28, 1963. Auguste remained at the institute in Zurich as professor of experimental physics until 1922, when he became a professor of physics at the Free University of Brussels (now two institutions, the Université Libre de Bruxelles and the Vrije Universiteit Brussel). His son Jacques was born the year he moved to Brussels. Piccard was keenly interested in studying cosmic rays in the upper atmosphere and designed a pressurized gondola that would carry a researcher to a high altitude without wearing a pressure suit. He made a total of 27 balloon ascents, reaching an altitude of 53,142 feet (16,197 m) on August 18, 1932.

In the 1930s Piccard became interested in deep-sea exploration and realized that the gondola he had designed for high-altitude research might be modified to withstand the pressures encountered in the ocean depths. The gondolas Piccard designed for his bathyscaphes were derived directly from those he had used for his balloon ascents. Piccard died in Lausanne on March 24, 1962.

THE *TRIESTE* AND ITS VOYAGE TO THE CHALLENGER DEEP

In 1958, having been purchased by the U.S. Navy for $25,000, the *Trieste* became the USS *Trieste* and was transported to the Naval Electronics Laboratory in San Diego, California, where it was prepared for its deepest dive, in an operation code-named Project Nekton. That preparation involved fitting the bathyscaphe with a new observation gondola. The replacement was made in three sections by the Krupp Steel Works at Essen, Germany. Its walls were made of steel five inches (12.7 cm) thick and designed to withstand an external pressure of 17,745 pounds per square inch (122 MPa). The gondola weighed 12.8 tons (13 t) in air and 7.9 tons (8 t) when submerged. Its single viewing window was the shape of a truncated cone, with the smaller end on the inside, and made from Plexiglas plastic, which was the strongest transparent material available. The original mercury-vapor arc lamps on the outside of the gondola were capable of withstanding tremendous water pressure.

The new gondola weighed more than the original, so the float had to be enlarged to support it. The new metal float was 59.5 feet (18 m) long, 11.5 feet (3.5 m) wide, and filled with 22,500 gallons (85 m³) of gasoline held in several tanks. The entire bathyscaphe, comprising both gondola and float, displaced 50 tons (45 t) of water when its gasoline tanks were empty and 150 tons (135 t) when they were full. The *Trieste* had a cruising speed of one knot (1.15 mph; 1.8 km/h).

On October 2, 1959, the refitted *Trieste* was hoisted on board the freighter *Santa Maria* and set sail for Guam in the western Pacific Ocean. On November 15, 1959, the ship left Guam and headed for the Mariana Trench (see "Mariana Trench" on pages 65–68). The *Trieste* was lowered into the water, and the USS *Wandank* towed it into position for its first dive. On that occasion the bathyscaphe dove to 18,600 feet (5,800 m), setting a new world record.

The *Trieste* dove into the Challenger Deep on January 23, 1960, carrying Jacques Piccard (1922–2008) and naval lieutenant Donald Walsh (1931–) to a depth of 35,798 feet (10,911 m). The two

The bathyscaphe *Trieste*, photographed on January 1, 1960, was a research submersible designed by Auguste Piccard (1884–1962) and his son Jacques (1922–2008). *(Central Press/Stringer)*

explorers climbed into the cramped gondola, opened the valves allowing water to flood into the air tanks in the float, and the craft began its descent beneath the waves at three feet per second (0.9 m/s). When they reached 27,000 feet (8,230 m), the crew released enough of the nine tons (8.2 t) of iron pellets in their ballast tanks to halve the rate of descent. The photograph shows the *Trieste* being loaded at Guam.

Four hours and 48 minutes after it sank beneath the surface, the *Trieste* settled onto the floor of the Challenger Deep. One of the outer panes of the Plexiglas window cracked during the descent, but the window did not fail, and the two men remained safe. When they switched on the exterior light, Piccard and Walsh saw a shrimp and a few soles and flounders—bottom-dwelling flatfish—and noted that the ocean floor was covered by *diatomaceous ooze*, which is a soft sediment consisting of more than 30 percent by volume of the silica shells of single-celled algae called diatoms. The presence of living animals proved that even at this depth the water contained dissolved oxygen.

They remained on the floor for 20 minutes, eating chocolate for energy. The temperature inside the gondola was 45°F (7°C). Outside the water temperature was 37.4°F (3°C). Then it was time to return to the surface, and they released more of the iron ballast. The ascent took three hours 17 minutes.

The dive of the *Trieste* set a record that is unique. Most records for speed or altitude can be bettered, but no one can ever dive deeper than the bottom of the Challenger Deep, because that is the deepest point on the Earth's surface. There is nowhere deeper, so the *Trieste* record can never be broken—and since its dive in 1960 no other manned attempt has been made to explore the bottom of the Mariana Trench.

In April 1963, after further modification, the *Trieste* took part in the search for the USS *Thresher,* a submarine that sank on April 10, 1963, during deep diving tests. *Trieste* found the submarine off the New England coast at a depth of 8,400 feet (2,560 m). The *Trieste* was then decommissioned and dismantled, but her original Italian-made gondola was fitted to *Trieste II,* a submersible that dove to the *Thresher* in 1964. The original *Trieste* is now displayed at the Naval Historical Center in Washington, D.C.

Jacques Ernest-Jean Piccard was born in Brussels, Belgium, on July 28, 1922. In 1943 he enrolled at the University of Geneva to study economics, but World War II intervened and he enlisted in the French army. After the war he taught at the university while also helping his father, Auguste Piccard, develop his bathyscaphe and demonstrate its potential for deep-water exploration. Following the *Trieste*'s dive in the Challenger Deep, Jacques and Auguste Piccard developed what they termed a mesoscaph, designed to carry tourists. During the Swiss National Exhibition of 1964–65 their first model, the *Auguste Piccard,* carried 33,000 passengers beneath the surface of Lake Geneva. In the 1970s Jacques Piccard established the Foundation for the Study and Preservation of Seas and Lakes and began warning about the dangers of pollution and overfishing. For the remainder of his life, he remained active in the design and development of submersibles. He died on November 1, 2008, at his home beside Lake Geneva, Switzerland.

Donald Walsh was born on November 2, 1931. He was educated at the U.S. Naval Academy, Annapolis, graduating in 1954 with a degree in engineering. He has a master's degree in political science from San

Diego State University and a master's degree and Ph.D in physical oceanography from Texas A&M University. Walsh spent 15 years at sea, retiring with the rank of captain. Most of his service was spent in submarines, and he commanded a submarine in the Pacific Fleet from 1968 to 1970. He was dean of marine programs and professor of ocean engineering at the University of Southern California, and in 1989 his company, International Maritime Incorporated, formed a joint venture with the P. P. Shirshov Institute of Oceanology to establish the Soyuz Marine Service, an underwater maintenance company, which continues to operate in Russia.

ALUMINAUT, ALVIN, AND THE DEEP-SEA SUBMERSIBLES

During World War II engineers began to consider the possibility of using aluminum rather than steel in the construction of submarines, and the concept of such a vessel was developed at Reynolds Metal Company in Richmond, Virginia, by Julian Louis Reynolds, the founder's son. Aluminum was used extensively in wartime aircraft construction, which probably required as much of the metal as the industry could produce at the time, and serious work on a submersible made from aluminum did not begin until after the war. In 1964 J. Louis Reynolds, by then executive vice president of Reynolds Metal, contracted the Electric Boat Division of General Dynamics, at Groton, Connecticut, to build the first aluminum vessel capable of diving deep into the ocean. The submersible, called the *Aluminaut,* was launched in September 1964.

Time magazine likened it to a "fatheaded sperm whale." It was certainly big. *Aluminaut* weighed 80 tons (73 t). It was 51 feet (15.5 m) long, and its cylindrical aluminum hull was 6.5 inches (16.5 cm) thick. That thickness of metal was designed to withstand pressure of 7,500 pounds per square inch (51.7 MPa), giving the submersible a maximum depth of 17,000 feet (5,180 m). It was sufficiently buoyant to be capable of submerging, ascending, and operating underwater entirely under its own power, and it had a cruising speed of 3.8 knots (4.4 mph; 7 km/h). It had external arms for collecting and manipulating specimens, and side scan sonar as well as active and passive sonar. *Aluminaut* had a range of 80 miles (130 km) and could remain submerged for up to 72 hours. It carried six people: a crew of three and three scientists.

The occupants had access to four windows through which to view their surroundings, and the vessel could carry up to 6,000 pounds (2,725 kg) of scientific equipment.

Aluminaut was based at Miami, Florida, and was operational from 1964 to 1970. It was owned and operated by Reynolds Marine Services, undertaking contract work for other organizations. It took part in submarine filming expeditions, including some led by the explorer, environmentalist, and filmmaker Jacques-Yves Cousteau (1910–97). The U.S. Navy hired *Aluminaut* to recover a torpedo lost from a testing facility in the Bahamas. The submersible also took part in surveys commissioned by the U.S. Naval Oceanographic Office, during which it dove to 6,000 feet (1,830 m).

The aluminum submersible's most famous mission was also on behalf of the military, and it had nothing to do with scientific research. On January 17, 1966, a B-52 bomber and KC-135 refueling plane collided in midair over Palomares, Spain. The bomber dropped four unarmed 1.45-megaton hydrogen bombs, three of them over land and one into the Mediterranean. The bombs on land were quickly located and recovered, but not the one that had fallen into the sea. As part of the 18-ship task force deployed in the search, the navy used its own submersible, the CURV (Cable-controlled Underwater Research Vessel), and also contracted the Reynolds *Aluminaut* and the *Alvin* (see below) to help in the search. The missing bomb was found on March 15, but when attempts were made to attach a line to it, the bomb slid away into deeper water. It was found again on April 2, and on April 7 *CURV I* recovered it.

The *Aluminaut* was retired in 1970. The Reynolds Metal Company donated it to the Science Museum of Virginia, in Richmond, Virginia, where it is now displayed.

It was *Alvin* that located the lost H-bomb. *Alvin,* known technically as a Deep Submergence Vessel (DSV), entered service on June 5, 1964. It is owned by the U.S. Navy and operated by the Woods Hole Oceanographic Institution (WHOI), at Woods Hole, Massachusetts. Its name honors Allyn C. Vine ("Al-Vin"; 1914–94), a physicist and physical oceanographer who was a senior scientist at WHOI and a tenacious proponent of the use of manned submersibles. He was unable to be present on *Alvin*'s first dive because at the time he was deep below the surface of the Atlantic Ocean in the French submersible *Archimède.*

Despite its age *Alvin* is still working, because over its lifetime it has been repeatedly modified and its equipment updated. By early July 2008 it had made more than 4,100 dives, carrying more than 8,000 researchers. It is designed to reach a depth of 14,764 feet (4,500 m), and it can reach the ocean floor over nearly 63 percent of the world's total ocean area. *Alvin* is 23.3 feet (7.1 m) long, 8.5 feet (2.6 m) wide, and 12.0 feet (3.7 m) high. It weighs 17.6 tons (17 t) and can carry a payload of 1,500 pounds (680 kg).

Alvin carries three people: a pilot and two scientists. They view their surroundings through three 12-inch (30-cm) windows, and there are powerful lights mounted outside the vessel, as well as film and video cameras. The submersible descends by flooding air tanks and rises when the pilot releases expendable steel weights. While *Alvin* remains near the bottom, the pilot can make the craft hover, rest on the bottom, or negotiate rugged terrain, and the vessel has two hydraulic arms that can manipulate specimens and lift 200 pounds (90 kg). Powered by five hydraulic thrusters, *Alvin* cruises at 0.5 knot (0.6 mph; 0.9 km/h) and has a top speed of two knots (2.3 mph; 3.7 km/h), a range submerged of three miles (5 km), and with three people on board its life-support systems allow it to remain submerged for 72 hours. Each dive lasts six to 10 hours.

In October 1968 *Alvin* suffered an accident while preparing for a dive off Cape Cod, and it would have been lost had *Aluminaut* not come to its rescue. *Alvin*'s support ship, the *Lulu*, was made from two naval pontoons linked by a support structure, with *Alvin* held by cables between the pontoons. One day a cable snapped while three people were on board, and the hatch was open. *Alvin* dropped into the sea and immediately sank. The three crew members escaped, but the submersible went down in 5,000 feet (1,500 m) of water. It remained on the seabed until September 1969, when *Aluminaut* found it and managed to secure a line to it, and *Alvin* was raised to the surface.

In 1974 *Alvin* took part in the French-American-Mid-Ocean Undersea Study (FAMOUS) alongside the French submersibles *Cyana* and *Archimède*. FAMOUS studied the Mid-Atlantic Ridge (see "Mid-Atlantic Ridge" on pages 58–62), finding evidence that helped confirm the theory of seafloor spreading. Soon after that study was completed, *Alvin* moved to the Pacific to examine the Galápagos Rift, where the seafloor is spreading even faster than it is in the Atlantic. In 1979 *Alvin* discovered black smokers—vents emitting hot water

Recovering the submersible *Alvin* after a dive. The *Alvin* is being lifted aboard its tender, the research vessel *Atlantis II*. *(Craig Dickson, Woods Hole Oceanographic Institution)*

and dissolved chemicals from below the seabed (see "Robert D. Ballard, Black Smokers, and Life at the Extremes" on pages 137–144).

All its many modifications have drastically altered the appearance of *Alvin* over the years. It also has a different support ship, the research vessel *Atlantis II.* The photograph shows the present version of *Alvin* emerging from the water after a dive and being lifted onto *Atlantis II.*

There are plans to retire *Alvin* and replace it with a submersible capable of diving to 21,320 feet (6,500 m), a depth that would give it access to more than 99 percent of the ocean floor. Until that craft is built, *Alvin* will continue to be the workhorse of the oceanographic community.

Life in the Abyss

Life on land depends on green plants. Plants harness sunlight as a source of energy that they use to synthesize carbohydrates from atmospheric carbon dioxide and water they obtain from the soil. Herbivorous animals feed on plants, carnivores feed on herbivores, and omnivores eat both plant and animal material. Marine life also depends on the aquatic equivalent of green plants. These are *algae*. Algae are simple plantlike organisms that perform *photosynthesis* but lack true roots, stems, and leaves; most are single-celled, but some grow large. Seaweeds are algae. Photosynthesis is the process by which green plants synthesize carbohydrates from atmospheric carbon dioxide and water absorbed from the soil, using light energy to drive the reactions.

Algae perform photosynthesis, which requires a source of light. This means they live only in the upper layers of the sea, where they drift with the tides and currents. Herbivorous animals feed on the algae, and other animals hunt the herbivores. Many of these organisms are microscopically small, and they are known collectively as *plankton*, a word derived from the Greek *plagktos*, meaning "wandering." The members of the plankton that perform photosynthesis are called *phytoplankton*; the Greek word *phutôn* means "plant." Most members of the phytoplankton are single-celled organisms with silica shells, called *diatoms*. The members of the plankton that feed on the phytoplankton are called *zooplankton*. Larger animals also feed on plankton, including the larvae of many fish species, and some very large animals, such as basking sharks *(Cetorhinus maximus),* which can grow to a length of more than 30 feet (9 m).

All of these organisms live near the ocean surface or in shallow water, where light can penetrate, and at one time it seemed entirely reasonable to scientists that nothing could inhabit the dark, cold depths. This view was first proposed in 1843 by Edward Forbes (1815–54) and came to be called the azoic hypothesis (see "Charles Wyville Thomson, Scientific Leader of the *Challenger* Expedition" on pages 48–52), but it did not go unchallenged.

This chapter tells of the oceanographic expeditions that first discovered animals living at great depths, thereby demolishing the azoic hypothesis. It then goes on to describe some of those animals and explains how they manage to survive in perpetual darkness and under huge water pressure.

Over almost all of their area, the deep oceans are cold. There are places, however, where very hot water gushes upward through fissures in the ocean floor, carrying with it a variety of dissolved chemicals. These places are called hydrothermal vents. The chapter describes them and the organisms that live around them—including some animals that are very strange indeed.

HMS *PORCUPINE* AND LIFE IN THE PORCUPINE ABYSS

On May 18, 1869, the 141-foot (43-m) wooden paddle steamer HMS *Porcupine* set sail from Valentia Island, off the southwestern coast of Ireland, to explore an area of the North Atlantic Ocean in the Bay of Biscay known as the Porcupine Bank, discovered during an earlier *Porcupine* expedition and named after the ship. On board, Charles Wyville Thomson (1830–82) and William Benjamin Carpenter (1813–85) were seeking evidence of life on the seabed.

Carpenter was one of the vice presidents of the Royal Society, a position that gave him some influence with the authorities, and he was able to persuade the Admiralty to equip the *Porcupine* with a dredge capable of reaching deeper than anyone had dredged before. The dredge is a traditional device used by fishing vessels to gather oysters and other bottom-dwelling shellfish. It consists of a large bag with a rectangular mouth that is fastened to an iron frame, the two long sides of the frame being in the shape of scrapers. Strong chains connect the frame to the thick rope used to lower, tow, and raise the dredge. When the dredge is dragged along the seafloor, the lower long edge of the frame scoops up seabed sediment and guides it into the bag. The finer sediment escapes through the weave of the bag, leaving behind all the animals, plants, and stones that were mixed with the sand and mud.

A dredge is very heavy, however, which makes it difficult to use in water deeper than about 20 fathoms (60 feet; 18 m). The dredge is much heavier to tow in deeper water, so the ship must be more powerful, and winching the dredge back to the surface by hand from a depth of 150 fathoms (900 feet; 274 m) is extremely hard work. The *Porcupine* was equipped with a 12 horsepower (9 kW) auxiliary engine (called a donkey engine) to assist in hauling up the dredge. This allowed the scientists to gather samples from deeper and deeper water, and on July 22, 1869, the dredge scooped up material from 2,435 fathoms (14,610 feet; 4,453 m). The dredge had been down for 7¼ hours, and its bag carried 170 pounds (75 kg) of material. In *The Depths of the Sea*, his account of the expedition published in 1873, Thomson wrote the following account:

> In 1869 we took two casts in depths greater than 2,000 fathoms. In both of these life was abundant; and with the deepest cast, 2,435 fathoms, off the mouth of the Bay of Biscay we took living, well-marked and characteristic examples of all the five invertebrate sub-kingdoms. And thus the question of the existence of abundant animal life at the bottom of the sea has been finally settled and for all depths, for there is no reason to suppose that the depth anywhere exceeds between three and four thousand fathoms; and if there be nothing in the conditions of a depth of 2,500 fathoms to prevent the full development of a varied Fauna, it is impossible to suppose that even an additional thousand fathoms would make any great difference.

Each time the dredge reached the surface with the first hauls, the scientists found that the bag contained little of any interest, but that many living animals were sticking to the outside of the bag and to the chains and the lower part of the rope. It seemed that the bag rapidly filled with mud, clogging the fabric and excluding the animals they were hoping to catch. Captain Culver, the commander of HMS *Porcupine*, devised a remedy. He had half a dozen swabs fixed to the dredge. Swabs were like birch brooms but made from hemp with long, hempen strands instead of wooden twigs. Sailors used them to dry the decks. Attached to the dredge the swabs trailed along the seabed, entangling animals that were then carried to the surface. Then they made a further modification by attaching a long iron bar to the bottom of the dredge bag and fastening large bunches of teased-out hemp to the ends of the bar. The device was very successful at catching animals with long legs or spines, but it seriously damaged the more fragile specimens.

The *Porcupine*'s dredge caught several species of mollusks, starfish, sea urchins, sponges, corals, worms, and vast quantities of foraminiferans—amoeba-like protozoans with a shell. They also caught fish from depths of 600–1,000 fathoms (3,600–6,000 feet; 1,100–1,830 m). These would have been greatly distorted as they entered a region of pressures much lower than those to which they were adapted. Thomson and Carpenter had shown convincingly that the ocean floor is not devoid of life.

GALATHEA AND THE PHILIPPINE TRENCH

In 1845 King Christian VIII of Denmark wrote to the Royal Danish Academy of Sciences and Letters informing its members that he had decided to send the corvette *Galathea* to the East Indies and especially to the Nicobar Islands, which were then a Danish colony. The expedition was to survey the natural products of the islands and assess their economic potential. The king instructed the academy to appoint "persons learned in the study of Nature and aides to assist them." The Galathea expedition was quickly organized, and the ship—43 feet (13 m) long and carrying 231 men, 36 guns, and supplies for one year—sailed around the world.

In the 1930s a group of Danish scientists and naval officers began planning a second Galathea expedition. They hoped it might leave in 1945, exactly 100 years after the original expedition, but World War II intervened, delaying the project, which finally took place from 1950 to 1952. The purpose of the expedition began to take shape in 1941, when the Danish author and journalist Hakon Mielche (1904–79) read a newspaper account of a lecture by the oceanographer and *ichthyologist* (a zoologist who specializes in *ichthyology*, the study of fish) Professor Anton Frederik Bruun (1901–61), in which Bruun entertained his audience with tales of sea snakes and said that, as a scientist, he would not dismiss the possibility that monsters resembling giant sea snakes might inhabit the deepest parts of the ocean. Mielche made contact with Bruun, and together they decided to organize an expedition to search for life in the deepest parts of the oceans.

Bruun and Mielche began preparations, and after a time they were joined by other Danish explorers and enthusiasts. They formed the Danish Expedition Foundation in June 1945 to raise funds for three expeditions. All three took place, but Galathea 2 was the biggest and scientifically the most important. Its principal aim was to verify Thomson's assertion that living animals thrived on the floor of the deepest oceans (see "HMS *Porcupine* and Life in the Porcupine Abyss" on pages 128–130). The expedition was well publicized and recorded, because Mielche arranged for journalists and a film crew to travel with it.

The *Galathea* was a frigate of the Danish navy, 260 feet (80 m) long, 36 feet (11 m) wide, with a crew of approximately 100 sailors and scientists. She sailed from Copenhagen on October 15, 1950, at first following in the steps of her predecessor, the first *Galathea.* Her route took her through the English Channel, southward along the African coast, around the Cape of Good Hope, northward to Mombasa, and across the Indian Ocean to the Nicobar Islands. From there she visited the Malayan Peninsula, Indonesia (formerly the East Indies), Australia, and New Zealand. She then crossed the Pacific to the United States, sailed southward along the coast, passed through the Panama Canal, and headed for home. *Galathea* arrived back at Copenhagen on June 29, 1952, where 20,000 people were waiting to greet her.

The first part of the voyage was spent visiting places where the original expedition had called a century earlier, to note the changes that had taken place in the intervening years. For much of 1951, however, the expedition concentrated on searching for living organisms in the deep ocean trenches beneath the Pacific Ocean. The deepest trench they visited was the Philippine Trench to the east of the Philippine Islands bordering the island of Mindanao, where the Eurasian Plate is sinking beneath the Philippine Plate. The Philippine Trench is about 820 miles (1,320 km) long and 19 miles (30 km) wide. It was first explored in 1927 by the German ship *Emden,* and in 1945 the USS *Cape Johnson* (see "Harry Hess, Robert Dietz, and Seafloor Spreading" on pages 68–72) measured its depth as 34,400 feet (10,497 m). At one time the bottom of the Philippine Trench was believed to be the deepest point on the Earth's surface; the Mariana Trench (see "Mariana Trench" on pages 65–68) is now known to be deeper.

Galathea used a bottom-trawl net. This is a traditional type of fishing gear consisting of a large net, closed at one end, with a wide mouth that is held open by floats, weights, and otter boards—curved plates resembling doors that pull to the sides as the trawl is towed forward. The otter boards disturb the seabed sediment, stirring it up into a cloud that hides the trawl net and makes a noise that attracts fish. The fish swim into the mouth of the net, and as the trawl continues to move forward the fish tire and drift toward the narrow, closed end (called the cod end), where they are trapped. On July 22, 1951, the *Galathea* crew raised a net filled with material trawled from the floor of the Philippine Trench at a depth of 33,433 feet (10,190 m). The bag contained sea anemones, shrimp, mussels, and sea cucumbers, all of which had been living in total darkness under a pressure of seven tons per square inch (1 t/cm^2). The shrimp species, about 1.2–2 inches (3–5 cm) long, has since been classified as *Galatheocaris abyssalis.* Biologists believe it lives close to the seafloor at depths of 14,765–16,400 feet (4,500–5,000 m), so it may have entered the trawl while the net was descending or rising. The two small white sea anemones were the most important find in this haul, because they reached the surface still attached to the rock on which they were living. Their existence proved that animals could live 1.5 miles (2.5 km) deeper than had been known until then. In other hauls the *Galathea* crew found a total of 115 different animal species living below 20,000 feet (6,000 m), including one fish trawled from 25,000 feet (7,500 m) in the Java Trench (also called the Sunda Double Trench), which extends for approximately 2,000 miles (3,200 km) in the eastern Indian Ocean.

The most astounding find, made by Dr. Henning M. Lenche (1904–77), was of several specimens of a *living fossil.* That is a member of a living species that is almost identical to a species otherwise known only from the fossil record, showing that it has evolved very slowly, changing little over many millions of years (see the sidebar "Living Fossils" on page 174). *Galathea's* living fossil, trawled in April

1952 from 9,800 feet (3,000 m) in the Middle America Trench off the coast of Costa Rica, was a mollusk about 1.4 inches (3.5 cm) across with a pointed shell, the apex projecting forward. It was given the name *Neopilina galatheae*. *Neopilina* belongs to the class Monoplacophora (the name means "with one plate"), and until the *Galathea* discovery scientists thought they had all been extinct for about 350 million years. The ancestors of monoplacophorans were probably flatworms, and *Neopilina* is believed to resemble closely the ancestor of all mollusks. More than 24 other monoplacophoran species have been discovered since *Neopolina* was found.

The second Galathea expedition was resoundingly successful and proved beyond any possibility of doubt that life exists on the ocean floor at every depth, down to the deepest places on Earth. Professor Bruun gave these deepest parts of the ocean the name *hadal*, from Hades, in Greek mythology the abode of the dead and the name of the god ruling it. That name is still used; the hadal zone is the region at the bottom of the deep ocean trenches. Bruun, Mielche, and their colleagues had confirmed Charles Wyville Thomson's prediction. On August 11, 2006, the third Galathea expedition with about 35 scientists sailed from Copenhagen on board the *Vædderen* on a nine-month voyage around the world, visiting Greenland, Africa, Australia, Antarctica, and South and North America.

INHABITANTS OF THE DARK DEPTHS

It was entirely reasonable for scientists to assume that no animal could survive the perpetual cold and darkness, and the immense pressure, of the deep ocean floor. The environment seemed utterly hostile, but its extreme hostility was not due to the temperature, darkness, or pressure. Animals can adapt to all of those, and most easily of all to extreme pressure. That is because the pressure inside a deep-sea animal is equal to that outside, so provided it remains at its accustomed depth the animal moves as freely and comfortably as a land animal that tolerates a pressure of 15 pounds on every square inch of its body (100 kPa) but maintains a similar internal pressure. What makes life on the ocean floor especially difficult is the shortage of food. The phytoplankton, on which all ocean life depends, is confined to the upper waters within reach of sunlight. No plants grow on the ocean floor. So what do the inhabitants of the deep ocean eat?

The answer to that question conjures a vision of a world that is very different from the world of animals that live in the sunlight and breathe air. Individual members of the plankton die, and so do the fish, mammals, and other organisms inhabiting the upper sunlit waters. On land the remains of dead organisms provide food for a hierarchy of scavengers and decomposers. There are scavengers in the ocean too, but they do not find their food lying still on the floor. Dead organic matter sinks through the water, and it is lost to them unless they are able to retrieve it before it disappears into the darkness below. In all, about one-fifth of the organic wastes and remains of dead organisms originating in the sunlit waters drifts all the way to the ocean floor. It falls like a thin but constant rain and accumulates as seabed sediment. On dry land the basic food—green plants—grows upward from the ground. On the ocean floor the basic food falls gently from above.

Some of the sediment returns to the surface off the coasts of continents, where the prevailing winds blow parallel to a coast that is aligned north and south, driving a current that flows approximately parallel to the coast and close to it. Because of the Earth's rotation the current flows at an angle to the wind, and the resulting forces produce a spiral effect in which the direction of the current changes with increasing depth. This motion slowly raises deeper water to the surface, bringing seabed sediment with it. The process is called *upwelling*, and it also occurs near tropical eastern coasts where trade winds blowing from the west push surface water away from the land, allowing deep water to rise, and sometimes in the open ocean when a vigorous storm such as a hurricane sets the surface water turning. The nutrients in the upwelling sediment greatly enrich the surface water, sustaining large populations of marine organisms.

Acorn worms (phylum Enteropneusta) are long wormlike animals that live in U-shaped tubes below the surface of the seabed sediment. Also living below the surface are copepods—crustaceans only 0.04 inch (1 mm) long—as well as still smaller eelworms (class Nematoda), water bears (phylum Tardigrada) measuring 0.001–0.4 inch (50–1,000 μm), and the minute larval stages of other organisms. Not all of the invertebrates are small, however. *Bathynomas giganteus* is closely related to pill bugs and woodlice (order Isopoda), but it is 15 inches (45 cm) long. Most mussel shrimps and seed shrimps (class Ostracoda) are about 0.04 inch (1 mm) long, but *Gigantocypris agassizi*, a deep-sea relative, grows to one inch (25 mm). There are many types of mollusk. Snails, limpets, whelks, cone-shells, and turret-shells are just a few of the gastropods that have been found in samples taken at depths of up to 15,750 feet (4,800 m) in the western North Atlantic, and squids and octopuses visit the ocean floor.

Animals living on or below the seafloor have abundant food, but food is scarce for fishes swimming clear of the ocean floor, 2,300–3,000 feet (700–1,000 m) below the surface. They must seize any food that comes their way, while avoiding predators. Many have large eyes, sensitive to low light levels, and many have organs called *photophores* that emit light—a property called bioluminescence (see the sidebar). The fishes use bioluminescence to attract prey, to produce a sudden flash of very bright light to deter or confuse a predator, and to locate mates. It is a risky strategy, of course, for light can also attract unwelcome attention, although some species use light as camouflage.

In order to take advantage of any opportunity of a meal, many predatory fishes are able to swallow prey that is larger than their own bodies. Gulper eels (11 species of *Saccopharynx*) hunt for fish at depths down to 6,000 feet (1,800 m) and are up to five feet (1.5 m) long. The mouth is huge, and as a gulper eel swallows its prey its stomach distends to accommodate it. Viperfish (*Chauliodus* species) spend the daytime about 15,000 feet (4,600 m) below the surface, but at night they rise nearer the surface, where food is more plentiful. They grow up to two feet (60 cm) long, with a slender body, a dorsal fin modified to a long rod called an *illicium* with a fleshy tip that acts as a lure, and a very large head with sharp teeth that are so long they will not fit inside the mouth. The jaws are hinged so that when the viperfish seizes prey it rotates its entire head upward and passes the victim back to a second set of teeth in the throat, called *pharyngeal jaws*. Its stomach can then expand to accommodate prey of any size that can pass through its jaws.

Black dragonfishes (three species of *Idiacanthus*) live 3,300–6,600 feet (1,000–2,000 m) below the surface. Females grow to about 15 inches (40 cm) long, but males to no more than two inches (5 cm). They have long slender bodies and large heads with big jaws and sharp fanglike teeth. The female has a *barbel*—a fleshy protuberance below the mouth—with a small bioluminescent tip. The larvae are long, slim, and transparent and have eyes on the ends of stalks that are up to half the length of the animal's body. Black dragonfish are highly bioluminescent. As well as the light on the barbel, they have photophores the full length of the upper and lower surfaces of their bodies and beneath their eyes. When disturbed they can suddenly illuminate their entire bodies.

Hatchetfish (family Sternoptychidae) have deep, laterally compressed bodies, with a narrow tail end that give them a shape reminiscent of a hatchet. They are covered in silver scales, and in some

BIOLUMINESCENCE

In the deepest parts of the ocean, where no daylight penetrates, intermittent flashes of tiny patches and rows of variously colored lights break what is otherwise total darkness. These lights are produced by living organisms. The production of light without heat by living organisms is called bioluminescence. It is not confined to deep-sea organisms: The most familiar examples are fireflies and fox fire (produced by fungi). Bioluminescence is especially common in the ocean, however, at all depths. Some members of the plankton are bioluminescent, and so are some corals, sponges, jellyfish, comb jellies, marine worms, crustaceans, squids, cuttlefish, fish, and others.

Deep-sea animals use bioluminescence for various purposes. The eyelight fish *(Photoblepharon palpebratus),* which is one of a group known as flashlight fishes, has a large luminescent organ beneath each eye, and it uses the light to help it see—as a flashlight. Anglerfishes have a fin ray that projects forward over the head, with a fleshy piece at the tip that acts as a lure to attract prey. In almost all the 100 or so species of deep-sea anglerfishes, the lure is luminescent. While the anglerfish remains motionless and invisible in the dark, it dangles its lure to attract a curious and unwary victim; as soon as the prey is within range the anglerfish seizes it. Some scaly dragonfishes (also called rat-trap fishes; family Stomiatidae) have organs near the eye that emit red light. This helps them to see the crustaceans that are their principal food; crustaceans cannot see red light, allowing the predator to approach undetected. Other deep-sea fishes use patches or rows of light to attract mates and to confuse and deter predators.

No vertebrate animal (fish, amphibian, reptile, bird, or mammal) is independently bioluminescent. Those that use light do so with the help of bioluminescent bacteria. The bioluminescent organs, called photophores, harbor colonies of bacteria that the fish supplies with nutrients and oxygen. The bacteria contain molecules of *luciferin*—five basic types of organic compounds, each found in a different group of organisms, which emit energy in the form of visible light. In the presence of oxygen and other substances (varying in different types of luciferin) and using the ATP $\leftrightarrow$ ADP (adenosine triphosphate $\leftrightarrow$ adenosine diphosphate) reaction as a source of energy, the enzyme *luciferase* catalyzes the oxidation of luciferin, generating reaction products that emit light.

species parts of the body are transparent. Their eyes are tubular and point permanently upward, suggesting that they hunt their prey by looking for fish that are silhouetted against the dim light above. Many fish hunt in this way, and the hatchetfish protect themselves from their own predators by means of photophores on the underside of their bodies. When the fish turns these on, they produce a faint light that is similar to the light penetrating from above, obscuring the outline of the hatchetfish. Hatchetfish spend the daytime at depths of about 12,000 feet (3,600 m) but rise close to the surface at night to feed on crustaceans. Different species range in size from 1.2 inches (3 cm) to two feet (60 cm).

Many of the fishes living below 6,500 feet (2,000 m) have very small eyes or are blind. This does not prevent the anglerfishes from exploiting bioluminescence to attract prey. Black seadevils (about nine *Melanocetus* species), which trap their prey about 10,000 feet (3,000 m) below the surface, are typical. Black in color, the females have a fin ray modified to form an illicium with a bioluminescent tip. The mouth is very large and set at an angle. The fish lies in wait for prey with its mouth wide open, ready to snap shut on any animal venturing within its reach. Females grow up to seven inches (18 cm) long and only they feed. The males are no more than one inch (2.5 cm) long, have no illicium, and do not feed. They live only to supply sperm but are free-living. In monster anglerfish (about 20 species in the family Linophrynidae) the mature males are tiny and remain permanently attached to the body of the female, living as parasites.

ROBERT D. BALLARD, BLACK SMOKERS, AND LIFE AT THE EXTREMES

Mid-ocean ridges are places where the oceanic crust is opening, pushing the seafloor away from the ridge on both sides and filling the resulting space with magma that cools to form new crust. That is how oceans grow wider (see "Harry Hess, Robert Dietz, and Seafloor Spreading" on pages 68–72). Geologists believe that the Red Sea is in an early stage of this process. It is widening as the African and Arabian tectonic plates move apart, and in time it will grow to the size of an ocean. At present the sea is about 1,240 miles (2,000 km) long and 17–224 miles (28–360 km) wide. A median trench runs along its center, and there the Red Sea is up to 1,600 feet (488 m) deep.

In 1949 an oceanographic survey discovered hot brines close to the seafloor near the Red Sea median trench, and in the 1960s further investigation revealed that the very salt water above the trench was at about 140°F (60°C) and below the water the seafloor mud was rich in metallic compounds. These discoveries indicated that water in the median trench was being heated from below. If that were so, then the same process should heat water above the mid-ocean ridges. In May 1976 scientists from the Scripps Institution of Oceanography were studying the Galápagos Rift, 8,200 feet (2,500 m) deep and running east-west between the Cocos and Nazca Plates, using a deep-tow vehicle equipped with instruments to measure water temperature and density, bottles for collecting water samples, acoustic sensors, and cameras. As the vehicle passed less than 130 feet (40 m) above the center of the rift, it detected a buoyant plume of hot water rising from a vent. When its photographs were developed and its water samples analyzed, the oceanographers had the first indisputable evidence of the existence of hot-water—hydrothermal—vents. The photograph shows a hydrothermal vent on a different part of the Mid-Atlantic Ridge. The location is different, but this is identical to the scene the cameras recorded at the Galápagos Rift. The scientists were looking for the first time at a black smoker.

Other photographs taken at the same time showed the ocean floor close to the hydrothermal vent, and to their surprise the scientists saw a variety of living animals including mussels, anemones, and crabs, as well as signs of burrowing. Far from being the barren desert they might have expected, cooked and poisoned by the hot sulfurous liquid gushing from the vent, there was a diverse and thriving community of organisms apparently finding nourishment from the chemical soup.

The following year scientists from the Woods Hole Oceanographic Institution (WHOI) examined the Galápagos Rift more closely, using the submersible *Alvin.* They found life was so abundant and so colorful that they named one vent the Rose Garden. Then, in 1979, another team arrived, this time led by Robert Ballard and Frederick Grassle, with a film crew from *National Geographic* on board the support vessel. The film they made was entitled *Dive to the Edge of Creation.* J. Frederick Grassle was then working at WHOI; he is now director of the Institute of Marine and Coastal Science at Rutgers University. Since 1974 Ballard, then a geologist and geophysicist at WHOI, had been head of the FAMOUS project (see *"Aluminaut,*

A hydrothermal vent on the Mid-Atlantic Ridge releases a sulfurous, mineral-rich fluid from a chimney rising 60 feet (18 m) above the ocean floor. The fluid, at a temperature of about 690°F (365°C), is black and looks like smoke, giving this type of vent its nickname of "black smoker." (The crew of the submersible *Alvin* took the photograph.) *(B. Murton/Southampton Oceanography Centre/Science Photo Library)*

Alvin, and the Deep-Sea Submersibles" on pages 122–126), and it was as part of that project that he and Grassle dove onto the Galápagos Rift. On the 1979 dive to 8,860 feet (2,700 m), Ballard and Grassle found several hydrothermal vents and a rich community of crabs, clams, and tubeworms living close to them. The animals were living in water at 46°F–60°F (8°C–16°C); the usual temperature at that depth is about 36°F (2°C).

Since that first discovery there have been many dives onto hydrothermal vents, and vent communities have aroused intense scientific interest. The fluids issuing from hydrothermal vents react with the seawater, and some of the reaction products solidify, forming chimneys that grow at up to 12 inches (30 cm) a day, with the vent fluids flowing up the center. Black smokers emit iron and sulfide that form iron monosulfide (FeS). This compound is black, and that is what gives the vent liquid its black color. As well as black smokers there are white smokers. These are cooler and flow more slowly than black smokers. They emit liquids rich in arsenic and zinc, and also compounds of barium, calcium, and silicon, which are white.

Vents remain open for only a few decades, and scientists have found old vents, now closed and cold, with dead animals lying around them. This suggests that vent organisms grow rapidly and have brief lives. Members of vent communities inhabit zones, each group of organisms living in water at a particular temperature. Tubeworms and mussels live closest to the vent, in the hottest water. Others live farther away, and animals at the edges are in water that is not much warmer than the deep-water average of 36°F (2°C). More recent expeditions found that a lava flow had covered the site of what had been the Rose Garden, but nearby the observers found a fresh community establishing itself on a site they named the Rosebud.

Most vent organisms are found nowhere else, although they are related to other marine animals, and some are exceedingly strange. *Riftia pachyptila* is one of about 100 species of tubeworms (also called beardworms). Tubeworms live inside tubes they secrete around themselves and feed by filtering particles from the water. Most are about 0.04 inch (1 mm) wide and up to 30 inches (75 cm) long. *R. pachyptila* is bigger. It grows up to almost eight feet (2.4 m) in length, but what makes it strange is that it has no mouth and no digestive system of any kind. It grows to a considerable size, but it never eats and is incapable of doing so. The photograph shows a colony of these animals, with some mussels and a crab, living beside a vent chimney in the Galápagos Rift. *Calyptogena magnifica* is a vent clam that grows up to eight inches (20 cm) long. It is able to feed by filtering particles, but that is not how it obtains most of its nutrients. Inside their bodies both the clam and the tubeworm have colonies of bacteria that obtain energy by oxidizing sulfides in the vent fluid and using the energy to synthesize organic matter from carbon dioxide. Some of the resulting nutrient passes into the tissues of the clams and tubeworms. The host animals keep the bacteria supplied with sulfides in the form of hydrogen sulfide (H_2S). Hydrogen sulfide blocks respiration, making it highly poisonous to most animals, but the worms and clams have a protein in their blood that binds to the H_2S, preventing it from causing harm, and releases it in the presence of the bacteria. Other vent organisms may feed in a similar fashion.

Giant tubeworms *(Riftia pachyptila)* living beside a hydrothermal vent on the Galápagos Rift *(D. Foster, WHOI/Visuals Unlimited)*

The white crabs in the picture may be feeding on colonies of bacteria growing on surfaces outside the vent.

Hydrothermal vents are extreme environments, but they are not the only places where organisms—mainly microbes—live under extreme conditions. There are organisms that thrive in hot springs, glaciers, salt evaporation ponds, and other apparently inhospitable places. Such organisms are now known as *extremophiles* (see the sidebar). When biologists began to examine the genetic composition of these organisms, they discovered that they did not fit into the structure that until then had been used to classify living organisms. This led to a dramatic revision of biological classification.

Robert Duane Ballard was born on June 30, 1942, in Wichita, Texas, and grew up in Pacific Beach, San Diego, California. He was educated at the University of California, Santa Barbara, graduating in 1965 in chemistry and geology, and in 1966 he earned a master's degree in geophysics at the University of Hawaii. He was called to active army service in 1967 and was transferred from the army to the navy as an oceanographer. He liaised between the Office of Naval Research and

EXTREMOPHILES

The biologists who first began classifying living organisms categorized all of them as either plants or animals, but by the middle of the 20th century it had become evident that fungi and many single-celled organisms fitted into neither of these categories. A more complex classification system developed, dividing organisms into five kingdoms: Bacteria, Protoctista, Animalia, Fungi, and Plantae. Then, in the 1970s, when biologists began comparing the DNA of different organisms, a team at the University of Illinois led by Dr. Carl Woese discovered that certain of the organisms considered to be bacteria were radically different genetically from true bacteria. Woese proposed calling this distinct group Archaebacteria, but later he abbreviated this to Archaea.

This regrouping led to a major revision of the classification of all living organisms. The classification now divides organisms into three domains: Bacteria, Archaea, and Eukarya (comprising the kingdoms Fungi, Plantae, and Animalia).

Archaea are single-celled organisms, but they are not closely related to Bacteria, and they include species that inhabit some of the most extreme environments on Earth. These species are known as extremophiles. Not all Archaea are extremophiles, but most extremophiles are members of the Archaea.

Acidophiles thrive where the pH is below 5.0, and some survive at pH 2.0 or even lower; the group includes representatives of the domains Archaea and Bacteria, and the kingdom Fungi.

Woods Hole Oceanographic Institution, and when he left the navy in 1970 he continued working at Woods Hole, where he devoted himself to raising funds for deep-sea exploration using the *Alvin*.

Ballard made his first deep-sea dive in 1969 in the submersible *Ben Franklin* off the Florida coast, and in 1970 he began mapping the Gulf of Maine for his doctoral dissertation. In 1974 he received a Ph.D. in marine geology and geophysics from the University of Rhode Island. He joined a French-American expedition in 1975, searching for hydrothermal vents over the Mid-Atlantic Ridge. This led to his work on the hydrothermal vents of the Galápagos Rift.

In 1985 Ballard sailed on the French research ship *Suroît*, which was searching for the wreck of the *Titanic*. The *Suroît* was recalled, and Ballard transferred to the *Knorr*, a Woods Hole research vessel working on a secret project funded by the U.S. Navy to search for two lost submarines. Once that work was completed, Ballard continued his search for the *Titanic*, which he finally located in September 1985 and studied in detail in 1986, using *Alvin*. Since then Ballard has led several expeditions searching for sunken ships.

Alkaliphiles thrive where the pH is above 9.0, in such places as soda lakes. The group includes members of the domains Archaea, Bacteria, and Eukarya.

Halophiles thrive in saline environments, requiring at least 2M (2 moles/liter) of salt (NaCl). Most belong to the Archaea, but there are some Bacteria and at least one member of the Eukarya, the alga *Dunaliella salina,* found in salt evaporation ponds (and used in cosmetics and dietary supplements).

Hyperthermophiles are members of the Archaea that thrive in temperatures of 176°F–250°F (80°C–121°C) and fail to reproduce at temperatures lower than 194°F (90°C); they were first discovered in the 1960s in hot springs in Yellowstone National Park.

Osmophiles thrive in environments with a very high sugar concentration; most are yeasts (kingdom Fungi).

Piezophiles (barophiles) thrive under pressures of 130 tons per square inch (38 MPa) or more; the bacterium *Halomonas salaria* (domain Bacteria) requires a pressure of 345 tons per square inch (100 MPa).

Psychrophiles (cryophiles) thrive in temperatures lower than 59°F (15°C), and many cannot live where the temperature rises above 68°F (20°C). Most are either Archaea or Bacteria. They occur in alpine and Arctic soils, and in glaciers, snowfields, and sea ice.

Thermophiles thrive in temperatures of 140°F–176°F (60°C–80°C); they are found in hot springs and hydrothermal vents.

In 2004 Ballard was appointed professor of oceanography at the University of Rhode Island. He is also director of the Institute of Archaeological Oceanography at the university's Graduate School of Oceanography, president of the Institute for Exploration in Mystic, Connecticut, and scientist emeritus at Woods Hole Oceanographic Institution.

MERMAID

Atargatis, a goddess in Assyrian mythology, was often depicted as part human and part fish. According to an Assyrian legend, Atargatis killed a human shepherd with whom she had fallen in love. Mortified with guilt and grief, she leaped into a lake intending to take the physical form of a fish, but the water refused to conceal her beauty, and while she acquired the lower body of a fish, her upper body remained unaltered. That story was written in about 1000 B.C.E., and it is believed to be the earliest account of a mermaid.

The first mermaid may have been Assyrian, but mermaid stories are told in every part of the world. They can talk with mortal men and have an unfortunate tendency to fall in love with them. This often leads to the drowning of the beloved, some say because mermaids easily forget that humans cannot breathe under water. In some cultures they are also omens of ill fortune, especially of storms, and were sometimes vindictive, luring sailors to their doom.

Mermen—human males above the waist and fish below—take little interest in human affairs, so they feature in fewer stories. In Greek mythology the son of Poseidon, god of the sea, and Amphitrite was Triton, who had the upper body of a man and the lower body of a fish. Triton was therefore a merman, and some stories tell of more than one Triton.

There are also references to mermen in medieval Scandinavian manuscripts. In *Speculum Regale* (The king's mirror), written in Norway in about 1250, the unknown author described a merman as follows:

> It is reported that the monster called merman is found in the seas of Greenland. This monster is tall and of great size and rises straight out of the water. It appears to have shoulders, neck and head, eyes and mouth, and nose and chin like those of a human being; but above the eyes and the eyebrows it looks more like a man with a peaked helmet on his head. It has shoulders like a man's but no hands. Its body apparently grows narrower from the shoulders down, so that the lower down it has been observed, the more slender it has seemed to be. But no one has ever seen how the lower end is shaped, whether it terminates in a fin like a fish or is pointed like a pole.

No one ever sees the merman at close quarters. The author also writes that no one has been close enough to see whether it has scales like a fish or skin like a man. This may give a clue to what the observers have seen—for clearly they have seen something, and there is another clue in the fact that mermen are usually seen shortly before a storm.

Some species of whales and dolphins engage in behavior called spy hopping, in which they rest vertically with their heads projecting above the water, perhaps in order to see what is going on above the surface. The underside of their bodies is paler than their backs, and seen from a distance this can give the appearance of a head on a narrower neck. Walruses, natives of the waters around Greenland, also sit upright in the water.

Surely, though, no sailor could mistake a whale, dolphin, or walrus for a man? Well, perhaps they might if a storm was imminent and warm air was moving over the top of a layer of much colder air about seven feet (2.2 m) thick next to the sea surface. Light is refracted as it passes across the boundary between the warm and cold air, and this can stretch images vertically. From a distance of about half a mile (1 km), the distortion makes a spy hopping animal appear much taller than it really is and narrower near the sea surface than above. It matches closely the Norwegian description.

In the early 19th century there was also fraud, and mermaids were a popular attraction at London fairs. In his book *Eccentricities of the Animal Creation,* published in 1869, the English antiquary John Timbs (1801–75) wrote that: "Less than half a century ago, a pretended Mermaid was one of the sights of a London season; to see which credulous persons rushed to pay half-crowns and shillings with a readiness which seemed to rebuke the record—that the existence of a Mermaid is an exploded fallacy of two centuries since."

Most of the supposed mermaids profitably displayed for the delight of Londoners were made by carefully attaching the tail of a fish to the upper body of a monkey and making sure that visitors were kept some distance from the attraction. The work was skillfully done and might well convince a person already prepared to believe in the possibility of mermaids. By the 1850s, however, all the exhibits had been pronounced fakes, and public interest waned. The belief in mermaids was not vanquished, however, and in the latter half of the century it revived, an interest triggered by the controversy surrounding

the 1859 publication of Charles Darwin's *On the Origin of Species by Means of Natural Selection.*

It seemed to some people that Darwin's evolutionary theory allowed for the evolution (and subsequent extinction) of a variety of mythical animals, but especially of mermaids. Darwin maintained that all vertebrate animals with true lungs had descended from an ancestor that possessed a swim-bladder, used to provide the neutral buoyancy that allows most fish to remain at a particular depth without expending energy. In 1860 Darwin wrote to his friend Charles Lyell: "*Our* ancestor was an animal which breathed water, had a swim-bladder, a great swimming tail" (italic in the original). Perhaps, then, mermaids had once existed as a missing link between fishes and humans. Opponents of Darwin's theory reversed the argument. Darwin apparently allowed the possibility that mermaids had once existed, but the idea of mermaids was clearly preposterous. Therefore, they argued, Darwin's idea was equally preposterous.

The argument quickly crossed the Atlantic. In 1861 the *Family Herald: A Domestic Magazine of Useful Information and Amusement,* an American weekly story paper that was published from 1843 to 1940, challenged anyone to demonstrate the link between fishes and humans. "Let them catch a mermaid," the paper demanded, "and they will find the missing link." In London the humorous magazine *Punch* was eager to join in the fun. In 1868 it published a cartoon of a mermaid holding a comb and mirror, with the caption "Mr. Punch's Designs from Nature (?) Toilette du Soir à la Sirène." It also reached the 1871 meeting of the British Association for the Advancement of Science, at which a Scottish anthropologist, Forbes Leslie, stated that he "had heard gentlemen quote the belief of intelligent persons, incapable of deception, who asserted that they had distinctly seen and watched mermaids." *Punch* reported this claim, adding its own suggestion that the mermaid (which it gave the taxonomic name *Siren canora*) was an evolutionary missing link. When HMS *Challenger* set sail on its 1872 expedition (see "HMS *Challenger*" on pages 44–45), *Punch* commented that ". . . not the least important results of the Expedition may be the acquisition of a Sea Serpent and the capture of a live Mermaid."

Eventually the debate moved on, and scientists became more interested in the more general possibility of composite organisms, possessing features of more than one group. Everyone accepted that mermaids were purely mythological beings. There are no such things as mermaids or mermen, nor can there be, because they are physiologically impossible.

SEA MONSTERS

Until the 18th century, when for the first time they were allowed to sail on naval hydrographical expeditions, few naturalists had any direct experience of the sea or of marine life, except for what they saw along shorelines. They were interested in the animals found in distant regions, but lacking any opportunity to visit those regions they had to rely on the descriptions of sailors and other travelers: on travelers' tales. The naturalists drew and described what had been described to them, but their depictions were at best approximate, and there was a great deal of confusion.

The walrus aroused intense interest. Few European naturalists ever visited northern Russia or Scandinavia where these animals might be found, but stories about them found their way south. Many of the illustrations of the walrus that were drawn in the 16th century were recognizable, but in 1635 the Spanish Jesuit Juan Eusebio Nieremberg (1595–1658) published *Historia naturae maxime peregrinae* (Natural history, most especially the foreign), a collection of descriptions of exotic beasts, most from the Americas, and containing a woodcut of an animal he called a *morss piscis*, which had a head resembling that of a dog, with protruding tongue, a mane, forelegs with long hair and large claws, and a fish's tail. The Swiss naturalist Conrad Gessner (1516–65) identified the morss piscis as the animal he had named *rosmarus*—now called the walrus, but with the taxonomic name *Odobenus rosmarus*.

The morss piscis was a real animal, albeit it barely recognizable. Many medieval animals were mythical, however. The serra, also called the sawfish and the flying fish, sounds as though it might be a real animal, but its description in a medieval bestiary—a book describing beasts—contradicts any such assumption. The three sentences that follow do not refer to any real animal:

> There is a monster in the sea called the serra, or flying-fish, which has huge wings. When the serra sees a ship in full sail on the sea, it raises its wings and tries to keep up with the ship for four or five

miles; but it cannot keep up the pace and folds its wings. The waves carry the wary creature back into their depths.

In the Middle Ages people believed that every aspect of the natural world had a moral meaning. The serra signified those people who start out performing good deeds but cannot keep it up and are overwhelmed with all kinds of vices, which drag them into the depths. "He that endureth to the end shall be saved" (Matthew 10:22). There are swordfish *(Xiphias gladius)*, of course, and also about 50 species of flying fish (family Exocoetidae), with enlarged and modified pectoral fins (and in some species also pelvic fins) that allow the fish to leap from the water to escape predators (including swordfish) and glide at up to 37 mph (60 km/h) for a considerable distance, sometimes flapping their "wings" to extend their flight. It is possible, therefore, that oft-repeated tales of flying fish being pursued by swordfish merged the two fishes into one, producing the serra.

The hydra was a more formidable beast. Conrad Gessner, said to be the greatest zoologist of his day, included it in his book on fish and aquatic animals *(De piscium et aquatilium animantum natura*; On the nature of fish and aquatic animals) published in 1558, but he was far from convinced that it really existed. The monster originated in Greek mythology and had no moral significance for Christians. The hydra resembled an aquatic serpent with many heads and poisonous breath. As one of the 12 labors he was required to perform as a penance, Hercules (originally called Heracles) was required to kill the hydra, which dwelt below water in a swamp, guarding an entrance to the underworld. Hercules covered his mouth and nose so he would not inhale its breath, then fired flaming arrows into its lair to draw it forth. When it appeared he attacked it with a sickle, but each time he cut off one of its heads two more grew in its place. So Hercules called on his nephew Iolaus to help him. He gave Iolaus a burning torch, and each time Hercules cut off a head Iolaus quickly burned the stump of the neck, preventing a replacement head from appearing. Hercules placed the hydra's one immortal head beneath a rock. Although it was mythological, the hydra may have been based on the octopus, which to this day is sometimes called the devilfish.

Hercules had other troubles. At one time a sea-nymph called Charybdis stole his cattle. Charybdis was the daughter of Poseidon and Gaia, and her misbehavior angered Zeus, who turned her into a sea monster. As a monster, Charybdis consisted mainly of an immense mouth that swallowed large amounts of seawater three times a day then forcibly ejected it again, causing whirlpools. Charybdis lived on one side of a narrow strait, and on the opposite side there was Scylla, another sea monster. Scylla had 12 legs resembling those of a dog, a cat's tail, and six necks each with a head containing three rows of teeth. She lived on a large rock, and sailors who tried to avoid one of these monsters were in danger of encountering the other. Their quandary may be the origin of the expression "between a rock and a hard place."

Monkfish is the name given to a number of species of North Atlantic fishes, some of which are caught commercially. In the 16th century there was also an animal called a sea monk that lived off the coast of Denmark. The Jesuit naturalist Caspar Schott (1608–66) included a picture of one in his book *Physica curiosa* (Physical curiosities), published in 1662, and the way he depicted it, with a human face and tonsure, it did look like a monk. Schott also drew a fish that looked like a bishop. In both cases he showed the fishes standing in an upright position, the sea monk on its tail fin and the sea bishop on two legs. In the 19th century the Danish zoologist Japetus Steenstrup (1813–97), professor of zoology at the University of Copenhagen and an authority on cephalopods—the class of mollusks that includes all squids, octopuses, cuttlefish, and the nautilus—reproduced a drawing of the sea monk beside a picture of a squid shown in a vertical position with its tentacles at the bottom. The two were fairly similar, and Steenstrup suggested that perhaps the sea monk was really a squid.

Sometimes real animals were mistaken for mythological ones. The narwhal *(Monodon monoceros)* is a member of the order Cetacea, which includes all whales, dolphins, and porpoises. It inhabits Arctic waters and is unique because in the male the upper left incisor tooth is extended forward as a long spiral tusk. Early accounts of the narwhal described it as a sea unicorn and depicted it as a fish with the head of a horse and its tusk set on its forehead, although by the 18th century it was being called a false unicorn.

LEVIATHAN

The leviathan was a biblical sea monster. In some of the stories about it, the leviathan was a huge fish that eats one whale every day, and according to one legend the whale that swallowed Jonah (Book of Jonah, chapters 1 and 2) narrowly avoided being swallowed by the leviathan. In the Jewish tradition, God originally created two leviathans, one male and one female, but then slew the female, lest their offspring should destroy the world. To Christians, the leviathan is associated with Satan.

Medieval sailors feared the leviathan. They believed the monster resembled a huge serpent that would swim rapidly round and round a ship until it made a whirlpool that drew the ship under the water, where the leviathan devoured it.

Scholars who have tried to identify an animal that might correspond to the leviathan have most often chosen the Nile crocodile *(Crocodylus niloticus)*, a reptile that occurs naturally in large rivers, lakes, and marshes throughout most of Africa, apart from the northwest and the Sahara. It is a large crocodile, up to about 16 feet (5 m) long, with a scaly body and fearsome teeth. It spends the night in the water and comes ashore shortly before dawn to spend the day basking in the warm sunshine. It does not need to feed every day, but when hungry it preys on large mammals that come to the water to drink. The crocodile seizes its prey, drags it into the water, holds it below the surface until it drowns, then tears off pieces of flesh by gripping the prey securely and spinning its own body. This may suggest the leviathan, but the fit is not perfect. Psalm 74 mentions the heads (plural) of the leviathan, and chapter 41 of the Book of Job has the leviathan breathing fire and smoke. So perhaps the leviathan was not a crocodile.

In *Moby-Dick,* published in 1851, Herman Melville (1819–91) refers to the white sperm whale of the title as the leviathan, and since then the name leviathan has come to be associated with the sperm whale *(Physeter catodon).* The sperm whale can grow to 60 feet (18 m) or more in length, so it is certainly big enough. Was it a whale that swallowed Jonah? All the Book of Job (1:17) says is that: "the Lord had prepared a great fish to swallow up Jonah. And Jonah was in the belly of the fish three days and three nights." Medieval writers did not believe that this was a whale. The anonymous author of a 13th-century bestiary held in the Bodleian Library at the University of Oxford is clear on the matter. "Whales are creatures of a huge size, which draw in and spout out water. They make greater waves than any other sea-creature." On the other hand, the author had no doubt about the identity of the beast that had swallowed Jonah. It was called the aspidochelon, and the author described it as follows:

> There is a monster in the sea which the Greeks call "aspidochelon"; the Latin "aspido" is a tortoise; it is also called sea-monster because its body is so huge. It was this creature that took up Jonah; its stomach was so great that it could be mistaken for hell, as Jonah himself said: "Out of the belly of hell cried I, and Thou heardest my voice" (Jonah 2:2). This creature raises its back above the waves, and it seems to stay in the same place. The winds blow sea-sand on it and it becomes a level place on which vegetation grows. Sailors believe it is an island, and beach their ships on it. Then they light fires, and when the creature feels the heat of the fire, it dives into the water and drags the ship down with it into the depths.

The author was wrong about the origin of the name; the Greek *chelōnē* means turtle, and the animal is also known as the asp-turtle, the first part of the name referring to the asp (Greek *aspis*), which used to be a general name for any venomous snake but most commonly for the Egyptian cobra *(Naja haje)*. A similar monster occurs in an Old English poem called "The Whale," where it is called fastitocalon.

This fabulous beast may have originated in stories recorded by the Roman naturalist Pliny the Elder (23 or 24–79 C.E.) in his *Naturalis historia* (Natural history), which he wrote in about the year 77, and the ancestor of all the medieval bestiaries was a book called *Physiologus,* written originally in Greek by an unknown author and translated into Latin in about 400 C.E. The book is called *Physiologus* because it begins many of its stories with the phrase "the physiologus says," physiologus usually being translated as the naturalist.

The physiologus drew the following moral lesson from the aspidochelon:

> The same will befall those who are full of unbelief and know nothing of the wiles of the devil, trusting in him and doing his work; they will be plunged into the fires of Gehenna with him.

There was a second lesson to be drawn from the manner in which the aspidochelon found its food, which the physiologus expounded as follows:

The nature of the monster is such that when it is hungry it opens its mouth, and gives out a sweet scent; the little fishes smell this and gather in its mouth. When the monster's mouth is full it closes its jaws and swallows them. The same will befall those who are not firm in their faith, and yield to all delights and temptations as if drunk with scents; and then the devil swallows them up.

There is one final candidate for a fish capable of swallowing a man. The goliath grouper *(Epinephelus itajara)*, a predatory fish found in shallow water along the coast of West Africa, in the western Atlantic from Florida to Brazil, and along the American Pacific coast from California to Peru has been known to stalk divers and attempt to ambush them. This fish grows to 8.2 feet (2.5 m) long and weighs up to 800 pounds (363 kg), and it is possible that even larger individuals exist, or have existed in the past. Groupers hunt by a quick rush and snap of the jaws, and they swallow most prey whole. They feed on crustaceans, fishes, octopus, and sea turtles, and there are stories of them grabbing divers. Was leviathan or the aspidochelon really a grouper? Who can tell?

SEA SERPENTS

Olaus Magnus was the Latinized name of Olaf Månsson (1490–1557), a Swedish priest, cartographer, and naturalist who fled Sweden to escape the Reformation there and eventually settled in Italy. In 1539 he published *Carta marina* (Map of the sea) in Venice—a map that was illustrated with small pictures, as was the fashion at the time, and many of his illustrations depicted ships being attacked by huge snakelike monsters. Olaus also wrote *Historia de gentibus septenrionalibus* (published in an English translation in 1658 with the title *History of the Northern People*), which described the daily lives of the peoples of northern Europe (the *septrionale* were the people who lived beneath the constellation of the Big Dipper or Plow). In this work Olaus described a sea serpent that lived on the coast of Norway near Bergen. He said it was 200 feet (61 m) long; 20 feet (6 m) thick; had two feet; long hair on its neck; sharp, black scales; and flaming red eyes. The monster lived in caves on the seabed, coming ashore on light summer nights to feed on calves, lambs, or pigs. At other times it fed out at sea on polyps, crabs, and other seafood.

There are many accounts of the Bergen sea serpent, extending over several centuries. Between 1752 and 1754 Erik Ludvigsen Pontoppidan (1698–1764), the Bishop of Bergen, published a two-volume work that appeared in English with the title *The First Attempt of the Natural History of Norway.* At first Pontoppidan doubted the tales he heard of the sea serpent, but the testimony of sailors and fishermen persuaded him. When asked whether sea serpents really existed, they considered the question as ridiculous as asking whether eels or cod really existed. Pontoppidan said that the serpents remained at the bottom of the sea except in July and August, when they spawned. Then they came to the surface in calm weather, but the slightest wave made them submerge.

Sea serpents were also reported from southern Greenland by Hans Egede (1686–1758), a Lutheran missionary, in his account of his experiences in Greenland in 1734. Egede wrote that on July 6, 1734, off the south coast of Greenland, he saw a sea monster that raised its head as high as the ship's yardarm. It had a long sharp snout; broad paws; and its rough uneven body was covered in scales. When it dived, the distance between its head and the tip of its tail appeared to be equal to the length of a ship.

One of the most widely quoted sightings of a sea serpent dates from 1848 and the British frigate HMS *Daedalus,* bound for Plymouth, England, after a tour of duty in the East Indies (now Indonesia). The story appeared in the *Times,* prompting officials at the Admiralty to ask for an explanation. In reply, Captain Peter M'Quhoe, the ship's commander, wrote the following letter to Admiral Sir W. H. Gage:

Sir, In reply to your letter of this day's date, requiring information as to the truth of a statement published in the *Times* newspaper, of a sea-serpent of extraordinary dimensions having been seen from Her Majesty's ship *Daedalus,* under my command, on her passage from the East Indies, I have the honour to acquaint you, for the information of my Lords Commissioners of the Admiralty, that at 5 o'clock P.M. on the 6th of August last, in latitude 24°44′S. and longitude 9°22′E., the weather dark and cloudy, wind fresh from the N.W., with a long ocean swell from the S.W., the ship on the port tack, heading N.E. by N., something very unusual was seen by Mr. Sartoris, midshipman, rapidly approaching the ship from before the beam. The circumstance was immediately reported by him to the officer of the watch, Lieutenant Edgar Drummond, with whom and Mr. William Barrett, the master, I was at the time walking the quarter-deck. The ship's company were at supper.

On our attention being called to the object, it was discovered to be an enormous serpent, with head and shoulders kept about four feet constantly above the surface of the sea; and as nearly as we could approximate by comparing it with the length of what our main top-sail-yard would show in the water, there was at the very least sixty feet of the animal *à fleur d'eau,* no portion of which was, in our perception, used in propelling it through the water, either by vertical or horizontal undulation. It passed rapidly, but so close under our lee quarter that had it been a man of my acquaintance I should have easily recognized his features with the naked eye; and it did not, either in approaching the ship or after it had passed our wake, deviate in the slightest degree from its course to the S.W., which it held on at the pace of from twelve to fifteen miles per hour, apparently on some determined purpose. The diameter of the serpent was about fifteen or sixteen inches behind the head, which was, without any doubt, that of a snake; and it was never, during the twenty minutes that it continued in sight of our glasses, once below the surface of the water; its colour, a dark brown with yellowish white about the throat. It had no fins, but something like the mane of a horse, or rather a bunch of sea-weed, washed about its back. It was seen by the quarter-master, the boatswain's mate, and the man at the wheel, in addition to myself and the officers above mentioned.

I am having a drawing of the serpent made from a sketch taken immediately after it was seen, which I hope to have ready for transmission to my Lords Commissioners of the Admiralty by to-morrow's post.

I have, &c.,

Peter M'Quhoe, *Capt.*

(*À fleur d'eau* means "awash.")

This report aroused much controversy. Some ship's captains who were familiar with that region of the Atlantic insisted that the monster had been a raft of giant kelp seaweed, but the *Daedalus* officers strongly disagreed, insisting that the thing was animate. Was it seaweed and, if it was, how was it moving at 12–15 mph (19–24 km/h)?

On July 30, 1915, this time off southwestern Ireland in the North Atlantic, the German submarine *U-28* sank the British ship *Iberian.* The ship sank quickly, and after about 25 seconds there was a loud explosion and wreckage and water was thrown high into the air. Mixed with the debris, there was also an animal resembling a crocodile, about 60 feet (18.3 m) long with four legs and webbed feet. The beast fell back to the surface, where it writhed for several seconds before disappearing below the water. If it was some kind of giant crocodile, what was it doing underwater off Ireland? And no known crocodile has webbed feet.

Sea serpent spottings also occur along the North American coast, and especially in Chesapeake Bay, in Maryland and Virginia, and in New England. The Chesapeake Bay serpent, called Chessie, has been reportedly seen at intervals for at least 200 years. It is usually described as a snakelike, limbless animal, about 30 to 40 feet (9–12 m) long, dark in color, with a head about the size and shape of an American football. Chessie has been seen countless times, and sometimes a sequence of observations by different people has tracked its movements. It has also been videotaped.

At about 7:30 P.M. on May 31, 1982, Robert Frew was at home with his family and friends. His house overlooked Chesapeake Bay, and everyone was in a part of the house with walls that were entirely glass, looking out over the water, when Chessie appeared about 100 feet (30 m) offshore. They estimated it to be about 30 feet (9 m) long with humps at intervals of about two feet (0.6 m). At first Mr. Frew watched through binoculars, but then he videotaped the object for about three minutes. At that time it was some 200 feet (60 m) from the house. Afterward several experts examined the videotape and said they did not think it a fraud. They passed it to the Smithsonian Institution, where a meeting of scientists was convened to consider it. On August 20 the Smithsonian meeting, chaired by George R. Zug, curator of amphibians and reptiles, examined the Frew tape as well as a color photograph of Chessie taken by Mrs. Kathryn Pennington on May 24, 1981, and sketches prepared by Clyde Taylor and his daughter Carol, who saw Chessie at about 8:20 P.M. on July 16, 1982. Six very experienced scientists studied the material. They concluded that the pictures were of an animal that they were unable to identify. One of those present at the meeting was Craig Phillips, of the Division of Hatcheries and Fishery Management Services of the Fish and

Wildlife Service. He dismissed the idea that Chessie might be an eel, sea snake, or a large snake such as a python or anaconda. Although these snakes enter water readily, they seldom enter seawater, require a water temperature of at least 70°F (21°C), and if they find themselves in the sea they immediately head for the shore. He tested this in Florida with a tame anaconda, releasing it into the surf; each time the snake made immediately for the shore.

Many reports of sea serpents are either hoaxes or simply mistaken. Currents can shape rafts of seaweed into a long, sinuous, perhaps serpentine form. That may explain the *Daedalus* sighting, despite the crew's denials. Squid swim with their tentacles trailing behind their bodies. Although they do not ordinarily swim close to the surface, if a large squid were to swim along the surface it could answer some of the descriptions of sea serpents—and experienced sailors would not expect to see a squid swimming in this way. A squid would not raise its head out of the water, however. Aquatic animals lift their heads out of the water only because they need to breathe air: They possess lungs, not gills. Porpoises sometimes swim in line, but their distinctive dorsal fins make sharks and dolphins instantly recognizable. It is also worth noting that most accounts of sea serpents have them undulating vertically, their backs forming a line of humps above the water. All snakes move their bodies horizontally, not vertically.

Do giant sea serpents exist? Despite the Frew tape, the countless anecdotal accounts, photographs, and drawings, no one knows. It is impossible to prove that they do not exist, but just one piece of evidence would prove conclusively that they do. Someone must produce a body, dead or alive, for scientific examination. Perhaps one day somebody will.

GIANT SQUID

It is not only sea serpents that inhabit the seas off Scandinavia. There is a much more terrifying monster: the kraken. Early accounts tell of a many-armed monster that could reach as high as a ship's main mast. Erik Pontoppidan (see "Sea Serpents" on pages 155–159) said it was undoubtedly the biggest of all sea monsters, 1.5 miles (2.4 km) wide and often mistaken for an island or even a series of islands. In later stories it is often described as a giant octopus that will attack ships and in some cases tip them over until they capsize and sink.

The most famous account of an alleged attack by a giant octopus appeared in *Histoire naturelle générale et particulière des mollusques,* a two-volume work by Pierre Dénys de Montfort (1766–1820), published in Paris in 1801–02. Dénys de Montfort was an expert on mollusks and worked at the Muséum national d'Histoire naturelle in Paris. According to his account, in 1783 an immense octopus had attacked a French ship sailing in the South Atlantic off Angola, Africa. The crew managed to free the vessel by chopping off the octopus's tentacles with axes and cutlasses. When they returned to Saint-Malo, in France, the sailors gave thanks in the chapel of St. Thomas, then commissioned a painting showing the incident, which was hung in the chapel. Dénys de Montfort claimed that his illustration was based on the Saint-Malo painting. The picture shows an octopus half out of the water and completely embracing a three-masted ship, with tentacles reaching to the top of all three masts, as well as around the bow and stern. It is not certain whether the Saint-Malo painting ever existed, and if the incident happened at all the Dénys de Montfort picture is a grotesque exaggeration. But this is not the only story.

On November 30, 1896, two boys, Herbert Cols and Dunham Coretter, were cycling along the beach of Anastasia Island, off St. Augustine, Florida, when they came across a large carcass half buried in the sand. They reported their find to DeWitt Webb (1840–1917), a local physician and naturalist, who examined it the next day. The carcass was 18 feet (5.5 m) long and seven feet (2 m) wide, and Webb estimated its weight as about five tons (4.5 t). It had what appeared to be long arms, and Webb thought it was an octopus. He informed Addison Emery Verrill (1871–1954), professor of zoology at Yale University and an authority on squids and octopuses. At first Verrill thought the remains were those of a squid, but then changed his mind and identified them as an octopus, which he named *Octopus giganteus.* Meanwhile, the tides washed the carcass off the shore and then stranded it again. When the entire mass was excavated, no arms were found, and when Webb sent Verrill samples of the tissue, Verrill declared it to be whale blubber with some skin attached, probably from a sperm whale.

Scientists accepted that what by then had become known as the St. Augustine Monster was a mass of blubber, a type of object called a *blob* or *globster.* Blobs are large masses of organic matter found on shores. They are often mistaken for the remains of sea monsters, but those that can be identified usually turn out to be the remains of known animals. It was not the end of the matter, however, because it transpired that samples from the blob had been sent to the Smithsonian Institution, which still held them. In 1971, 1986, 1995, and finally in 2004 different scientists examined them and analyzed their composition. Some thought they were from an octopus, but eventually the biologists decided that the blob was, indeed, from a whale.

Giant octopuses do exist. The biggest known species is the North Pacific giant octopus *(Enteroctopus dofleini)*, which can grow to a span of about 20 feet (6 m), measured from tentacle tip to tentacle tip. Octopuses live on the sea bottom, sheltering in small caves and rock crevices, so divers seldom encounter them. When they do appear, octopuses are usually harmless. If provoked, however, they may attack, and a giant octopus is very strong and capable of leaping upon its opponent. Divers have been drowned by octopuses that gripped them firmly and prevented them surfacing.

It is very unlikely that a giant octopus would rise to the surface and make an unprovoked attack on a ship. There is, however, a denizen of the deep that fits the description of the kraken—albeit a monster much smaller than Pontoppidan claimed—but it is not an octopus and, ironically, the first real evidence of its existence came from studies of sperm whales, the source of many blobs.

The sperm whale *(Physeter catodon)* is the largest toothed animal in the world and the largest carnivore. Adult males can grow to a length of 67 feet (20.5 m). They are able to dive more than 7,000 feet (2,100 m) below the surface and to remain submerged for as long as 90 minutes. Sperm whales feed on fish and squid, both of which they hunt at great depths. Today they are protected, but at one time whalers used to hunt them. When they were hauled onto the whaling ships, their skin was often covered in circular scars. These were made by the suckers of the squid that the whales hunt, but the size of the scars suggested a squid much larger than those found near the surface.

Squid are *cephalopod* mollusks, closely related to octopuses. Octopuses have eight tentacles, squid eight arms and two long tentacles. Both the arms and tentacles are lined with suckers. The squid uses its tentacles to seize prey and bring it within reach of the arms that surround its mouth. These pass it to the sharp, horny beak that tears the prey to pieces. There are 298 known species of squid. Most are no more than 24 inches (60 cm) long when adult and far too small to fight back if attacked by a sperm whale, yet the scars on sperm whales were undoubtedly made by squid, and squid beaks were often found in the stomachs of these whales.

Giant squid, big enough to fight sperm whales, have occasionally washed up on shore and sailors have seen them. They are classified as the genus *Architeuthis,* and there may be up to 20 species. There have been many sightings of them. In 1861 a French corvette fired rifles and cannon at a squid the crew estimated to be 25 feet (7.6 m) long, but it escaped. In 1922 a squid with arms 50 feet (15 m) long and one remaining tentacle 100 feet (30 m) long washed ashore at Port Simpson, British Columbia, Canada. This specimen, like others that had washed up with the tide, had a hook 12 inches (30 cm) long and 10 inches (25 cm) wide at the tip of its tentacle.

Perhaps the most interesting sighting occurred in the 1930s, when three giant squid attacked the *Brunswick,* a 15,000-ton (13,620-t) auxiliary tanker of the Norwegian navy that was sailing between Hawaii and Samoa. The squid, estimated to be about 30 feet (9 m) long not counting the tentacles, surfaced astern of the ship, which was sailing at about 12 knots (13.8 mph; 22.2 km/h). The squid caught up with the ship, swimming parallel to it at a distance of about 100 feet (30 m). When each squid was level with the bow, it turned and slammed hard into the side of the ship and tried to grasp the ship with its tentacles. Unable to get a grip on the smooth metal, each squid in turn slid back until it fell into the ship's propellers and was killed. This incident took place in full daylight, and the ship's captain, Commander Arne Grønningsœter, watched it from the bridge.

These squid deliberately attacked the ship, just as sailors had been claiming for centuries. Probably the squid mistook the ship for a whale. Until then biologists had thought that sperm whales hunted giant squid and that the squid might defend themselves. The *Brunswick* incident showed that sometimes the squid attacked whales, and a squid would have a considerable advantage in such an attack. A squid swims backwards by jet propulsion, drawing water into its mantle, closing the mantle, and forcing the water out through a narrow tube. This allows it to move very fast. It can also hover absolutely still in the water, invisible in the darkness and, because it is an invertebrate and soft-bodied, the whale's echolocation (see "Charles Bonnycastle and the Dream of Charting the Ocean Floor" on pages 52–54) may not detect it. In 1965 the crew of a Soviet whaling ship watched a sperm whale and giant squid fighting near the surface. Eventually they killed each other.

Giant squid females are larger than males. The largest females are estimated to be 43 feet (13 m) long, and the largest males 33 feet (10 m), including the tentacles. Some may reach a larger size, but those

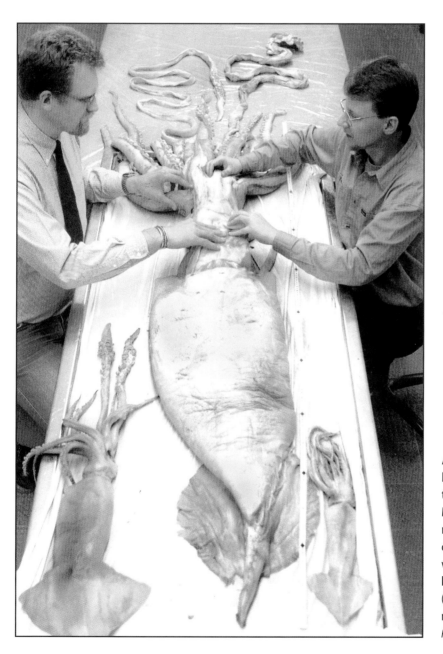

A giant squid (*Architeuthis*) being examined by scientists at the German Oceanic Museum in Stralsund, Germany. This specimen was caught in New Zealand waters—it is 20 feet (6 m) long and weighs 110 pounds (50 kg). Giant squid are often much bigger than this. *(Jens Koehler/AFP/Getty Images)*

that attacked the *Brunswick* were probably smaller than reported. The illustration shows a 20-foot (6-m) giant squid being examined by scientists.

Big though it is, the giant squid is not the largest of all squid. There is also the colossal squid *(Mesonychoteuthis hamiltoni)*. It was first described in 1925, based on body parts found in the stomach of a sperm whale. Colossal squid are heavier than giant squid but no longer. The largest so far discovered was a female, caught alive in 2007 by New Zealand fishermen. It was 14 feet (4.2 m) long, including the tentacles, although the tentacles had shrunk considerably after it died, so it would have been longer when it was first caught.

OARFISH, SUNFISH, AND MEGAMOUTH SHARK

Genuine living animals may account for at least some sightings of sea monsters. Oarfish, for example, have sometimes been mistaken for sea serpents. There are four species of oarfish, but the one most likely to be mistaken for a serpent is *Regalecus glesne,* also called king of the herrings. It is the longest fish in the world, with reports of individuals at least 56 feet (17 m) long, although most specimens are smaller, growing up to about 26 feet (8 m). Its long body is flattened, tapering toward the tail, and generally eel-like. The silvery body bears black or gray markings but with a highly variable pattern. Its shape, somewhat reminiscent of an oar, may account for its common name, but this may also refer to its pelvic fins, which consist of long spines ending in wider sections. The rays of its dorsal fin are pink or red and form a crest running the length of the body, and the much longer rays over its head can be erected, as in the photograph. As the illustration shows, it is a spectacular fish.

Oarfish are distributed throughout tropical and temperate seas. It lives at depths of 65–655 feet (20–200 m) and feeds on small fish and invertebrate animals. The oarfish seldom surfaces, but if it should do so close to a boat, it would be an extraordinary sight and might easily be mistaken for a serpent. Only one oarfish has been filmed alive in the sea, in 2001, by a group of United States sailors who were inspecting a buoy. They saw that the fish swims by undulating its fins while keeping its body straight. It has also been seen swimming in a vertical position.

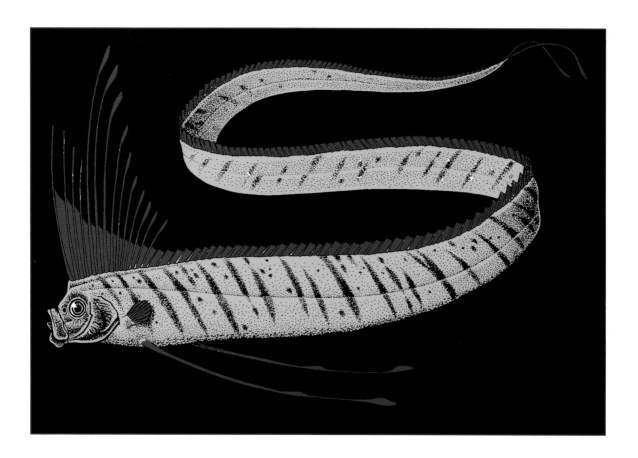

Another candidate for monster sightings is the ocean sunfish *(Mola mola)*. It is the world's heaviest fish, with an average adult weight of 1.1 tons (1 t), although some reach 2.4 tons (2.2 t). Sunfish can be 10 feet (3 m) long and measure 14 feet (4.3 m) from the top of the dorsal fin to the bottom of the anal fin. They inhabit all temperate and tropical seas, but they are seldom seen because they spend most of the time at depths of 650–2,300 feet (200–700 m). Their diet consists largely of jellyfish, augmented by crustaceans, squid, and other invertebrates. The photograph shows a sunfish swimming in the North Pacific.

Sunfish are harmless, but they are capable of giving people a fright. In August 2005 the three members of the Grey family were fishing for lobsters off the coast of Pembrokeshire, Wales, in their 14-foot (4.3-m) boat when they saw the triangular dorsal fin of a sunfish

The oarfish *(Regalecus glesne)* is sometimes called the king of the herrings. It is the longest of all bony fish, some reaching 56 feet (17 m). *(Richard Ellis/Photo Researchers, Inc.)*

The sunfish *(Mola mola)* is the world's heaviest fish, weighing an average 1.1 tons (1 t). It inhabits tropical and temperate seas throughout the world. This one is swimming off San Diego, California. *(Gregory Ochocki/ Photo Researchers, Inc.)*

projecting above the water. They approached closer for a better look, but the sunfish vanished. Then it leapt from the water and onto the boat, landing on top of four-year-old Byron, who fortunately escaped with only some bruises. This sunfish was small, weighing only an estimated 110 pounds (50 kg).

The sunfish is shaped like a huge plate, and although it usually swims in a vertical position, in warm weather sunfish sometimes lie on their sides at the surface, presumably basking to raise their temperatures. Its unusual shape and large size, together with the fact that it seldom rises to the surface and is less common in temperate waters than it is in the Tropics, ensure that any appearance will attract interest. Unfamiliarity and a partial view of the animal may well lead an observer to describe it as an unknown monster.

On November 15, 1976, a U.S. naval research vessel, *AFB-14*, was working about 26 miles (42 km) off Kaneohe, Hawaii, and had deployed two large parachutes as sea anchors at a depth of about 540 feet (165 m). The ocean in that area is 15,100 feet (4,600 m) deep. When the parachutes were hauled on board the crew found a fish, 14.63 feet (4.46 m) long, entangled in one of them. The sailors realized that the fish was unusual and took it ashore to be studied by scientists. It was 1983 before zoologists published a full description of the fish, which they recognized as a species that had never before been seen. They called it the megamouth shark *(Megachasma pelagios)*.

The shark owes its name to its huge mouth, positioned at the very front of its head—unlike most sharks, in which the mouth is on the underside of the head. The megamouth shark has thick, soft, rubbery lips, and its jaws carry many small hook-shaped teeth. The shark has two unequal-sized dorsal fins and a *heterocercal tail*—having a very large upper section and a smaller lower section. It feeds on krill (a type of shrimp), jellyfish, and zooplankton, which it catches by swimming slowly forward with its mouth wide open. The first five specimens to be found were all dead, but the sixth was alive, and scientists were able to track its movements for two days. They found that it spent the day about 490 feet (149 m) below the surface, rising to about 56 feet (17 m) at night. Krill also move nearer the surface at night, so the shark was probably following them. The photograph is of the shark that scientists were able to follow.

It was November 1984 before a second megamouth shark was caught, this time by anglers fishing with a deep-sea net off Catalina island, California. That individual was 14.73 feet (4.49 m) long. It is now preserved in the Los Angeles County Museum. A male, 16.9 feet (5.15 m) long, was washed ashore about 30 miles (50 km) south of Perth, Western Australia, in August 1988; it is displayed in the West Australian Museum. In all, only about 40 megamouth sharks have ever been caught. The largest, found in Sagami Bay, in Kanagawa prefecture,

The megamouth shark *(Megachasma pelagios)* grows up to 18 feet (5.5 m) long and weighs up to 1.3 tons (1.2 t). *(Fox Shark Foundation)*

Japan, in May 2006, was a female 18.7 feet (5.7 m) long. Megamouth sharks are now known to inhabit the tropical Atlantic and Pacific Oceans. They have not yet been encountered in the Indian Ocean, but probably they also live there.

The megamouth shark is a true giant, but there are two other species of very large sharks. The basking shark *(Cetorhinus maximus)* and whale shark *(Rhincodon typus)* both feed on plankton, which they filter from the

water by swimming slowly with their huge mouths wide open. The basking shark is found throughout the world wherever the water temperature is 46°F–57°F (8°C–14°C). It often swims close to the shore, sometimes in shoals, so it is a fairly familiar sight. Most basking sharks are 20–26 feet (6–8 m) long, but they can reach a length of more than 40 feet (12 m). The whale shark is the largest of all fish, although the oarfish is the longest. The largest recorded specimen of whale shark, caught in 1947 near Karachi, Pakistan, measured 41.5 feet (12.65 m). Individuals up to 59 feet (18 m) long have often been reported, but none of the reports has been verified.

Whales are the largest marine animals, and the blue whale (*Balaenoptera musculus*) is possibly the largest animal ever to have lived on Earth. An adult male may be 110 feet (33 m) long and weigh 200 tons (182 t). Blue whales feed exclusively on plankton, especially krill—of which an adult consumes up to four tons (3.6 t) every day.

SEA SNAKES

Sea serpents do exist, at least in the literal sense. There are 17 genera with 62 species of aquatic snakes. Two species live in freshwater lakes, one *(Hydrophis semperi)* in the Philippines and the other *(Laticauda crockeri)* on Rennell Island, one of the Solomon Islands. All the other species live in the sea. They occur in coastal waters throughout the tropical Indian and western Pacific Oceans, with some found around the coasts of Oceania. Some enter rivers and have been seen 100 miles (160 km) inland.

They are true snakes that have fully adapted to an aquatic life, and they are fairly helpless on land, where they often writhe desperately and strike at anything nearby. All sea snakes have paddle-shaped tails, and many have a laterally compressed body. These features help them to swim. The exceptions are the five species of sea kraits (*Laticauda* species), which spend part of their time on land, although they feed only at sea, most species on moray and conger eels, but some on fish, squid, and crabs. Sea kraits have only slightly flattened tips to their tails. The illustration shows *Laticauda laticauda*, a sea krait found in the sea around New Caledonia. Its flattened tail helps it to swim.

Being reptiles, not fish, sea snakes possess lungs rather than gills, and so they must surface to breathe, but they do not need to do so very often. Sea snakes can remain submerged for several hours and

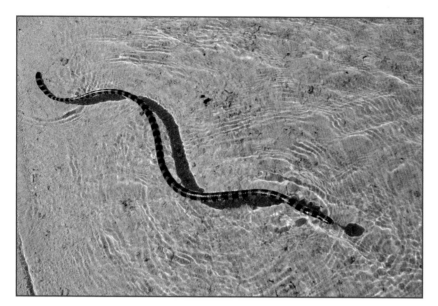

A sea krait *(Laticauda laticauda)* swimming in the sea off New Caledonia. Sea kraits are unusual among sea snakes in that they breed on land. This snake grows to about 3 feet (1 m) long. *(Daniel Heuclin/Photo Researchers, Inc.)*

have been observed swimming at depths of more than 295 feet (90 m). Their ability to remain below the surface is enhanced by the fact that they absorb oxygen through their skin. Seawater contains dissolved oxygen—it is what fish breathe through their gills—and sea snakes utilize it. The yellowbelly or pelagic sea snake *(Pelamis platurus)* obtains 20 percent of its oxygen in this way.

Most sea snakes feed on fish, especially eels. Some species that live on coral reefs have thin necks and small heads, which allow them to pull small eels out of their hiding places. Other snakes feed on crustaceans, and one species specializes in eating fish eggs. All sea snakes are highly venomous. They are descended from the Elapidae—the family of snakes that includes the cobras, kraits, mambas, coral snakes, and taipans. Most are not dangerous to people, however, because they are not aggressive and when they do bite they usually inject only a small amount of venom. They sometimes become trapped in fishing nets; the fishermen free them and throw them back into the sea and are rarely bitten. Snakes such as the yellowbelly sea snake that seize their prey and swallow it at once use their venom for defense. These species may bite if they feel threatened. Others, such as the sea kraits, that use their venom to immobilize prey, are less likely to bite. All sea snakes should be treated with caution, however, and some species and individuals of almost any species may make an unprovoked attack.

Sea snakes are usually four to five feet (1.2–1.5 m) long. The biggest is the yellow sea snake *(Hydrophis spiralis),* which grows up to nine feet (2.75 m) long. Snakes are often seen basking at the sea surface early and late in the day, but they are very obviously snakes and not monsters. Occasionally, however, they congregate in large numbers. In the early 1930s passengers on a steamer in the Strait of Malacca between the Malay Peninsula and Sumatra saw what they estimated to be millions of orange and black snakes twisted together into a long line that was 10 feet (3 m) wide. The ship followed the snakes for about 60 miles (96 km).

Perhaps a mass of sea snakes might be mistaken for a sea serpent, but there are also genuinely large snakes: the boa constrictor, reticulated python, and anaconda. The largest recorded boa constrictor *(Boa constrictor)* was 18 feet (5.5 m) long. It is reluctant to enter water, however, and does not swim in the sea. The reticulated python *(Python reticulates)* grows to about 33 feet (10 m). It enters rivers and ponds but not saltwater. The three species of anaconda—dark-spotted *(Eunectes deschauenseei)*; green *(E. murinus)*; and yellow *(E. notaeus)*—are semiaquatic and spend much of their time in the water. These massive snakes need the buoyancy provided by the water to help support their bulk. The longest anacondas are probably about 23 feet (7 m) in length. There are many tales of much bigger individuals, but they have not been verified. In 1948, for example, Amazonian Indians are alleged to have killed a 130-foot (40-m) anaconda. Although they are semiaquatic, however, anacondas rarely enter seawater. It is unlikely that any of these snakes could be mistaken for a sea monster.

COELACANTH

The oceans are so vast, so deep, and so inaccessible that they have ample room to hide many secrets. These include animals that may not be monsters, in the sense of being frightening, but that are certainly strange. On December 21, 1938, the trawler *Nerine* was fishing off the South African coast not far from the mouth of the Chalumna River. When the trawl net was hauled on board with its catch of small sharks, rays, starfish, and rat-tailed fish, one of the fish that spilled onto the deck was pale blue and about five feet (1.5 m) long. The fish lay on the deck for several hours, during which time it attracted the attention of the captain, Hendrik Goosen, by snapping at his hand.

Goosen thought the fish was likely to be inedible, but it also appeared unusual, so rather than throwing it back into the sea he decided to take it ashore and offer it to Marjorie Courtney-Latimer (1907–2004), curator of the East London Museum, who sometimes bought interesting-looking fish for the museum's collection.

The *Nerine* returned to port in East London, South Africa, and the message about the strange fish reached Courtney-Latimer. It was a hot day, she was busy, and the only reason she decided to go down to the dock was to wish the *Nerine*'s crew a merry Christmas. When she saw the fish, however, she thought it the most beautiful she had ever seen. She bought it and persuaded a taxi driver to allow her to take it back to the museum in his cab. The museum had no access to facilities for refrigerating the fish, so Courtney-Latimer had a taxidermist try to preserve it.

Courtney-Latimer could find nothing like this strange fish in any of the museum's reference books. It was clearly special, and so, on December 23, she sent a sketch of it to her friend Professor James Leonard Brierley Smith (1897–1968), with the following letter:

> Dear Dr Smith
>
> I had the most queer looking specimen brought to notice yesterday. The Capt of The Trawler told me about it so I immediately set off to see the specimen which I had removed to our Taxidermist as soon as I could. I however have drawn a very rough sketch and am in hopes that you may be able to assist me in classing it. It was trawled off Chulmna coast at about 40 fathoms.
>
> It is coated in heavy scales, almost armour like, the fins resemble limbs, and are scaled right up to a fringe of filment (*sic*). The Spinous dorsal, has tiny white spines down each filment. Note drawing inked in red.
>
> I would be so pleased if you could let me Know what you think, though I know just how difficult it is from a discription (*sic*) of this kind.
>
> Wishing you all happiness for the season.
> Yours sincerely,
> M. Courtney-Latimer

Professor Smith was a distinguished ichthyologist. From the written description and sketch he recognized at once that this fish was very remarkable indeed. The key phrase in Courtney-Latimer's letter was "the fins resemble limbs." This indicated a lobe-finned fish belonging to the subclass Crossopterygii. Crossopterygians are bony fish in which all the fins except for the tail fin are on movable stalks or lobes. The only difficulty with this identification was that, so far as scientists knew, the entire subclass of crossopterygians became extinct about 80 million years ago.

When he received the letter on January 3, 1939, Smith was at his vacation home, and for the moment there was nothing he could do about it. The most important thing, from his point of view, was to remain exceedingly cautious, lest he become a laughingstock. He telegraphed Courtney-Latimer asking her to keep the fish's internal organs, simply saying that the fish was "interesting." It was February 16 before he and his wife, also an ichthyologist, were able to get to

The coelacanth (Latimeria chalumnae) is a living fossil. Its closest relatives became extinct about 80 million years ago. (Peter Scoones/ Photo Researchers, Inc.)

East London. Courtney-Latimer was out when he arrived, and the caretaker showed them into the room containing the fish. Smith said later that he shook with excitement when he saw it. Then, as he examined it in great detail, he confirmed his first impression. The animal was, without the slightest doubt, a coelacanth. He and his wife worked together on a scientific paper announcing the discovery. This was published in June 1939. Smith named the species *Latimeria chalumnae*, in honor of Marjorie Courtney-Latimer and the Chalumna River, where the fish had been found.

The coelacanth was a living fossil—a species whose closest relatives have been extinct for many millions of years (see sidebar). The photograph shows a live coelacanth. Unfortunately, the internal organs of the East London specimen had not preserved well, and Smith needed another specimen. Supposing that coelacanths lived in or near the Mozambique Channel between Africa and Madagascar, Smith had posters distributed, written in English, French, and Portuguese, offering a reward to anyone who could supply him with one. In December 1952 a second specimen was caught in the Comoro Islands just to the north of the Mozambique Channel, and the South African prime minister authorized the use of a military aircraft to take Smith to the Comoros to collect it. Local fishermen were quite familiar with the fish, which they called gombessa, but had no interest in it because it was inedible—as Captain Goosen had correctly surmised.

Other coelacanths were later caught around the Comoro Islands, and for a time the South African specimen was thought to have been a stray from the Comoros population. More recently, however, more coelacanths have been discovered in South African waters, where they have been studied by deep-sea divers and by scientists using a submersible. Coelacanths are also known to inhabit waters off the west coast of Madagascar, and off Mozambique and Kenya, and in 1997 and again in 1998 coelacanths were captured off northern Sulawesi, Indonesia. Analysis of the DNA from tissue samples revealed marked differences between the Indonesian and African populations, and in April 1999 the Indonesian coelacanths were classed as a different species, *L. menadoensis*.

Coelacanths live in the twilight zone, 500–800 feet (150–245 m) off the steep slopes of volcanic islands, but often swim in shallower water. In 1991, a team of scientists led by ichthyologists Hans Fricke of the Max Planck Institute in Göttingen, Germany, and Phil Heemstra of the J. L. B. Smith Institute in Grahamstown, South Africa, using the German submersible *Jago,* found a group of coelacanths inside a cave. The fish emerge at night to feed. Their diet consists of fish

LIVING FOSSILS

Occasionally, an animal or plant species is discovered that bears a strikingly close resemblance to an ancestral species that became extinct millions of years ago. The extinct species are known only from fossils, and the living species that closely resemble them are described as living fossils, a term that was introduced by Charles Darwin (1809–82) in the first (1859) edition of *On the Origin of Species by Means of Natural Selection.* Darwin used the term informally, and to this day it has no agreed definition. The central concept is of a species or higher grouping that has changed very little over millions or hundreds of millions of years. Most species do not survive for long enough to qualify for this description. Mammal species exist for an average 1 million years before becoming extinct, and bird species for about 2 million years. For this reason the term *living fossil* is most often applied to higher taxa, especially to genera or families. Some genera of marine invertebrates survive for 100–150 million years.

Calling an organism a living fossil does not mean that it is identical to its fossil relative, implying that no evolution has taken place over a very long period, but only that the living fossil has evolved at a much slower rate than have other organisms. There are several examples of living fossil animals. These include the tuatara, coelacanth, nautilus, and horseshoe crabs.

The two species of tuatara *(Sphenodon punctatus and S. guntheri)* are mainly nocturnal reptiles resembling lizards that live on Stephens Island and a few other islands off the coast of New Zealand. The tuatara's skull is less evolutionarily advanced than that of the lizards and snakes. It is the only living representative of the Sphenodontidae, a family whose other members became extinct about 70 million years ago.

The two species of coelacanth *(Latimeria chalumnae and L. menadoensis)* belong to an order of fishes (Coelacanthiformes) whose other members became extinct about 80 million years ago. It is a bony fish that retains certain features resembling those of sharks, rays, and skates as well as features from still earlier fishes.

The subclass Nautiloidea contains four or possibly five species of *Nautilus* and the related *Allonautilus scrobiculatus.* Nautiloids are cephalopods, related to octopuses and squid, but differing from them in possessing an external shell. They are the only cephalopods with external shells to survive the mass extinction that occurred about 65 million years ago, and the nautilus shell has changed very little over the past 200 million years.

Horseshoe crabs, also called king crabs *(Limulus polyphemus,* two species of *Carcinoscorpius,* and *Tachypleus tridentatus)* first appeared in the Silurian period (443.7–416 million years ago), and they have changed little over the past 250 million years. They are arthropods that live on the seabed, feeding on other invertebrates, but they come ashore to breed.

including eels, skates, and sharks, as well as octopuses, squid, and cuttlefish, and they find prey by swimming slowly just above the seabed, although they can swim quickly if startled. Sometimes a coelacanth performs a headstand, keeping its snout on the seafloor. The largest coelacanth so far discovered was a female six feet (1.8 m) long, weighing almost 200 pounds (90 kg).

Far from harassing sailors, coelacanths are considered to be endangered, and they are protected under the Convention on International Trade in Endangered Species (CITES). Trade in coelacanths is illegal, and any that are caught alive must be released. Professor Smith had warned that unrestricted hunting of coelacanths might threaten their survival, and he proposed that an international society be formed to protect them. In 1987, Mike Bruton, director of the J. L. B. Smith Institute, together with other ichthyologists, founded the Coelacanth Conservation Council, based in Moroni, capital of the Comoro Islands.

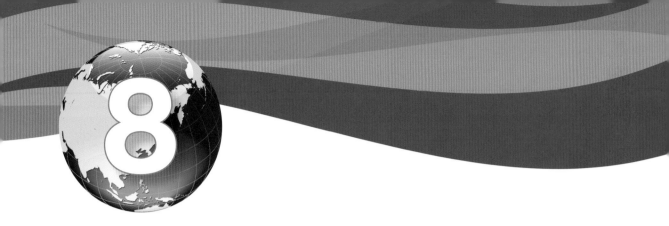

Modern Explorers

Ocean exploration continues apace as we enter the 21st century, backed by technologies far more powerful than were available to scientists at any time in the past. Satellites are able to measure sea levels and to monitor changing amounts of sea ice, transmitting a constant stream of data to observers in ground stations. Weather satellites monitor cloud formations and, from that, allow meteorologists to plot the development and movement of weather systems. Meteorological information is shared globally, and national weather services receive warnings of potentially hazardous conditions long before they arrive.

Robots are also acquiring data about ocean waters. Until recently, scientists relied on commercial and military seagoing vessels to provide them with information about seawater temperature and salinity and variations in the speed and direction of ocean currents. This system provided reliable data, but it was limited and left scientists ignorant of conditions over the large areas ships seldom visit. Most commercial ships travel along well-established shipping lanes, and shipboard measurements of seawater temperature are taken only at the level of the engine cooling-water intakes. That has now changed. In 2000 the first Argo floats were released in a cooperative program by 23 countries.

Named after the ship that in Greek mythology carried Jason and his crew of Argonauts in search of the Golden Fleece, the Argo floats spend most of their time below the surface, drifting with the tides and currents. At intervals, usually every 10 days, each float rises slowly to the surface, taking about six hours to do so. As it ascends, the float measures the water temperature and salinity. When it reaches the surface, the float makes radio contact with an orbiting satellite, which notes its position and receives its data. The float then sinks again, to drift until it is time to transmit more measurements. Comparing the position of each float with its previous position allows oceanographers to track changes in ocean currents and the temperature and salinity readings record changes in the ocean waters. By July 2008 all 3,218 of the planned Argo floats had been released. They are now distributed throughout every ocean, providing scientists with constantly updated measurements at varying depths of waters that ships never visit. Jason, leader of the Argonauts, is also commemorated. Jason is the name of the satellite that uses radar altimetry to monitor sea level with an accuracy of 1.3 inches (3.3 cm).

Below the oceans, scientists continue to study the ocean floor. They follow in the tradition of HMS *Challenger* (see "HMS *Challenger*" on pages 44–45), but now they can drill deep into the seabed sediment. The samples they return to the surface provide a record of the history of the Earth over many thousands of years. This final chapter describes a few of these present programs to explore the oceans.

DEEP-SEA DRILLING

Submersibles are able to visit the deepest ocean floor, allowing scientists to observe life in places that were once thought to be barren. Equipped with external arms, submersibles can scoop up samples of sediment. They can hover close to hydrothermal vents. What they cannot do, however, is study the seabed beneath the surface sediments. To do that requires drilling equipment so heavy that the operation can be carried out only from a large ship on the surface.

In the years following the end of World War II technological advances meant that the deep ocean floor began to come within the reach of oceanographers. The first attempt to drill deep into the ocean floor began in 1961 with Project Mohole, led by the American Miscellaneous Society and funded by the National Science Foundation. Mohole aimed to drill all the way through the oceanic crust to the Mohorovicic discontinuity marking the boundary between the

Earth's crust and mantle. The project acquired the *CUSS 1*, an oil-drilling ship capable of working in water up to 600 feet (183 m) deep. The team modified the vessel to work in deeper water, and Project Mohole drilled five holes off the coast of Guadalupe, Mexico. The deepest penetrated 601 feet (183 m) into the seabed, in water 11,700 feet (3,570 m) deep. Eventually Project Mohole ran out of funding, and it was abandoned in 1966.

In 1968 the first ship capable of drilling into the ocean floor was launched at Orange, Texas. That ship was called the *Glomar Challenger* (see sidebar), and for 30 months it cruised the oceans, drilling cores from the seabed. The operation, called the Deep Sea Drilling Project, was based at the Scripps Institution of Oceanography at the University of California, San Diego. It was supported by the National Science Foundation and planned by an advisory group of 250 scientists from all over the world, working under the auspices of the Joint Oceanographic Institutions for Deep Earth Sampling (JOIDES). JOIDES is a consortium of 31 U.S. institutions engaged in oceanographic research.

The *Glomar Challenger* scientists made several important discoveries about the ocean floor and continental drift. In order to succeed, however, the geologists and engineers had to solve a number of technical problems. No one had ever before attempted to drill deep into the ocean floor, working from a ship at sea. The task involved working with very long *drill strings*—the pipes that extend from the upper collar on the ship's deck to the cutting bit at the bottom of the drill. As the bit cuts into the rock below the ocean floor, the workers on deck lower cylinders down the hole. These fill with material from the drill hole, and when each cylinder is full the deck workers winch it back to the surface and lower another. Core cylinders pass up and down the drill pipe, but when the drill bit wears out, replacing it involves raising the entire drill string. Then, when the new bit has been fitted, the string must be returned to the same hole, otherwise drilling could never proceed beyond the depth attainable before a bit needs replacing. The *Glomar Challenger* team perfected a technique for achieving this in 1970, by placing a reentry cone 16 feet (4.88 m) in diameter into the hole before retracting the string, and using sonar scanning equipment to guide the string back into the hole. The technique developed for guiding the drill string back into the hole was then adapted to allow instruments to be lowered into drill holes that would take geophysical and geochemical measurements during and

GLOMAR CHALLENGER

The first phase of the Deep Sea Drilling Project (DSDP), based at the Scripps Institution of Oceanography, University of California at San Diego, began on June 24, 1966, when contracts were signed for the construction of a new drilling vessel, to be called the DV *Glomar Challenger*. Construction of the ship began on October 18, 1967, by the Levington Shipbuilding Company, in Orange, Texas. The ship was launched on March 23, 1968, and the Scripps Institution accepted it for service on August 11.

The ship's entry into service marked the start of the second phase of the DSDP, during which *Glomar Challenger* worked for about 30 months in the Atlantic, Pacific, and Indian Oceans, and in the Red and Mediterranean Seas, drilling cores from the seabed sediments. Each core was 30 feet (9 m) long and 2.5 inches (6 cm) in diameter. The cores are now stored at the Lamont Doherty Earth Observatory at Columbia University, New York, and at the Scripps Institution of Oceanography,

University of California, San Diego. That phase of the project ended in August 1972.

On Leg 3 of their journey, cores that the *Glomar Challenger* crew drilled along the Mid-Atlantic Ridge in the South Atlantic provided definite proof for continental drift and the production of new oceanic crust at rift zones. The cores also revealed that nowhere is the ocean floor more than 200 million years old.

By the end of the DSDP *Glomar Challenger* had sailed 375,632 nautical miles (431,977 miles; 695,051 km). It had acquired a total of 19,119 seabed core samples in water up to 23,111 feet (7,044 m) deep. It had drilled 5,712 feet (1,741 m) below the seabed and 3,543 feet (1,080 m) into the basalt rock of the oceanic crust.

Once the second phase of the DSDP had ended, *Glomar Challenger* continued to be used in scientific and engineering research. Its service with the DSDP ended in November 1983. The ship was then scrapped.

after the drilling operation. Some holes were equipped with instruments to monitor seismic events.

The DSDP ended in 1983 and was replaced by the Ocean Drilling Program (ODP), which ran until 2003. The National Science Foundation funded the ODP, and scientists from 22 nations took part in it. The ODP operated under the auspices of JOIDES and was conducted from Texas A&M and Columbia Universities. In 2003 the Integrated Ocean Drilling Program (IODP) replaced the ODP. The IODP, which is scheduled to continue at least until 2013, greatly expands on the work of the DSDP and ODP. It is funded by the National Science Foundation; the Japanese Ministry of Education, Culture, Sports, Science and Technology; the European Consortium for Ocean Research Drilling; the People's Republic of China Ministry of Science and Technology; and the Interim Asian Consortium, which is represented by the Korea Institute of Geoscience and Mineral Resources. The program studies life in the deep oceans and seafloor sediments in order to improve understanding of environmental changes and processes and natural cycles in the oceanic crust.

After *Glomar Challenger* was retired, its place was taken by the *JOIDES Resolution*, which has been joined by the *Chikyu*, a Japanese vessel whose name means "Planet Earth." *JOIDES Resolution* entered service in 1978 as an oil exploration vessel with the name *Sedco/BP 471*. It was converted for scientific research, and in January 1985, with a new name that refers to Captain James Cook's ship (see "James Cook, the Greatest Chart-Maker" on pages 15–21) it began working for the ODP. On a typical operation the ship carries a crew of 65 and about 50 scientists and technicians. The illustration shows *JOIDES Resolution* at sea. The ship is 469 feet (143 m) long, 68.9 feet (21 m) wide, and the top of its drilling derrick stands 202 feet (61.5 m) above the sea surface. It carries more than 5.5 miles (9 km) of drill pipe and can drill in water up to 27,000 feet (8,230 m) deep. The drill string leaves the ship through the moon pool, a hole 23 feet (7 m) wide from the deck to the sea below the ship. *JOIDES Resolution* worked continuously until 2005, when it was taken out of service to be upgraded. Following sea trials in 2007, it resumed work in 2008. *Chikyu* was launched on January 18, 2002, and entered service on July 29, 2005. It is 690 feet (210 m) long, 125 feet (38 m) wide, and carries 100 crew and 50 scientists. Its drill string is 6.4 miles feet (10 km) long, and its derrick rises 230 feet (70.1 m) above the sea. Its moon pool measures 39 feet by 72 feet (12 m by 22 m).

By examining cores of sediment drilled from the ocean floor, scientists can trace the history of the Earth's climate and the ways in which climate changes have affected marine life. Studies of the crustal rocks beneath the layers of sediment record the formation and movement of tectonic plates. Drilling into the ocean floor yields information about the structure and history of the planet that can be acquired in no other way.

INTERNATIONAL COUNCIL FOR THE EXPLORATION OF THE SEAS

Drilling programs explore the physical world. The biological world is no less important. Modern fisheries, equipped with large, powerful vessels and advanced technologies for locating and catching fish, have depleted stocks of many commercially important fish populations. Pollution of the seas, from ships, from the air, and from substances washed from the land and carried to the coast by rivers, is of widespread concern.

In the 1980s a series of meetings were held under United Nations auspices. Together they comprised the United Nations Conference on the Law of the Sea, and their most notable achievement, in 1982, was the establishment of economic exclusion zones (EEZ). These now have the status of international law. Every nation is entitled to control all entry to and every activity within the sea along its coasts but must allow ships of all nations to move freely in waters more than 12 miles (20 km) from the coast. The EEZ extends 200 miles (322 km) from the coast—a distance based on the average distance from a continental coast to the edge of the continental shelf—and within it the coastal nation has a right to control all commercial activities and exploitation. (Where the sea between two countries is less than 200 miles [322 km] wide, the governments agree to a midway EEZ boundary.) It would be unlawful, for instance, for an oil company based in one country to drill for oil in the EEZ of another country without that country's agreement.

No nation owns or has exclusive rights in the seas beyond its EEZ. A variety of intergovernmental organizations negotiate agreements on conservation and environmental protection measures in international waters. Marine research is also coordinated and promoted by international bodies. In the North Atlantic region—including the Baltic Sea and North Sea—the International Council for the Exploration of the Seas (ICES) is the responsible organization.

The ICES was founded in 1902, and its headquarters are in Copenhagen, Denmark. It has 20 member countries: Belgium, Canada, Denmark, Estonia, Finland, France, Germany, Iceland, Ireland, Latvia, Lithuania, the Netherlands, Norway, Poland, Portugal, Russia, Spain, Sweden, the United Kingdom, and the United States. Australia, Chile, Greece, New Zealand, Peru, and South Africa are affiliates, and the Worldwide Fund for Nature (also called the World Wildlife Fund in North America) and BirdLife International have the formal status of observers. More than 1,600 scientists from the member countries exchange information, principally on marine life and ecology, through the ICES. Its eight scientific committees specialize in

JOIDES Resolution is the research ship that was built in the 1970s for JOIDES (Joint Oceanographic Institutions for Deep Earth Sampling). It carries more than 5.5 miles (9 km) of drill pipe, allowing it to take samples from deep below the ocean floor. *(Ocean Drilling Program, Texas A&M University)*

fisheries technology, oceanography, resource management, marine habitat, mariculture, living resources, the Baltic Sea, and diadromous fish species (diadromous fish spend part of their life cycles at sea and part in freshwater). The ICES uses the information gathered from its scientists to prepare advice that it offers to the governments of its members. The ICES publishes reports and other publications as well as, through Oxford University Press, the *ICES Journal of Marine Science,* which is a scientific journal.

Modern scientific research is a collaborative endeavor. The days of the solitary investigator working in private are long gone—if they ever truly existed. Today scientists work as teams, usually with members drawn from different institutions that may be located in any part of the world. International institutions help coordinate their work and hold conferences and meetings at which individual scientists can meet and exchange ideas. The ICES performs this function for the North Atlantic. The Mediterranean Science Commission (CIESM), based in Monaco and with 23 member countries, is the equivalent organization in the Mediterranean and Black Seas. The Intergovernmental Oceanographic Commission, with the slogan "One Planet, One Ocean," is part of UNESCO (the United Nations Educational, Scientific, and Cultural Organization), and its responsibilities extend to all oceans, seas, and coasts.

Conclusion

Today, after many centuries of exploration and scientific investigation, the oceans that cover more than two-thirds of the Earth's surface still conceal many mysteries. Scientists know, for instance, that the way the oceans absorb and store solar radiation and the way ocean currents transport away warmth from the equator strongly influence the world's climates, but many details of that influence remain unknown. Until they are revealed, long-term climate predictions can never be anything but very approximate.

Biologists know, broadly, the relationships among living organisms in the oceans, but they have much still to learn. This is especially true of microorganisms. The usual way to identify bacteria involves growing colonies in laboratory cultures in order to separate individual species and study their DNA. Biologists are able to cultivate less than 1 percent of all the bacteria found in seawater, however. Seawater also abounds in viruses, about which even less is known. Yet marine bacteria may produce compounds of great medical value, and they have other abilities that would be of use to society if they could be harnessed. For one thing, there are marine bacteria that feed on oil. If cultures of them were stockpiled, they might be used to break down oil slicks resulting from leaks and spillages, without causing additional environmental harm.

Much of the ocean floor remains to be studied in detail, and it, too, may conceal resources. Hard, approximately spherical nodules rich in metals were first discovered on the ocean floor in 1803.

Often called manganese nodules—manganese is the most abundant metal found in them—they are formed by bacterial action and in a world of shortages might one day be a useful source of manganese. In high latitudes, sedimentary seabed rock contains methane—the principal ingredient of natural gas—locked inside the crystal structure of ice. These methane hydrates contain possibly twice as much hydrocarbon fuel as all of the world's fossil fuel stocks—coal, oil, and gas—combined.

This book has provided a brief summary of the history of ocean exploration and discovery, illustrated as it were by snapshots, each recording a particular scientist or team and a particular discovery. Exploration continues, and at an accelerating pace. The story has not ended.

GLOSSARY

abyssal   the ocean floor between 6,500–13,000 feet (2,000–4,000 m) and 20,000 feet (6,000 m) depth.

abyssalpelagic   the zone of the open ocean lying between 6,500–13,000 feet (2,000–4,000 m) and 20,000 feet (6,000 m) depth.

acidophile   an **extremophile** that thrives where the pH is below 5.0.

algae   simple, plantlike organisms that perform **photosynthesis** but lack true roots, stems, and leaves; most are single-celled, but some grow large. Seaweeds are algae.

alkaliphile   an **extremophile** that thrives in very alkaline conditions, typically where the pH is greater than 9.0.

aphotic   the region of the ocean below about 650 feet (200 m), where no sunlight penetrates.

asthenosphere   the weak region of the Earth's upper mantle, 60–120 miles (100–200 km) below the surface.

azimuthal   a plane map projection that plots features from a central point onto a plane surface tangent to the sphere.

azoic   containing no trace of life.

barbel   a fleshy protuberance near the mouth of a fish.

bark (barque)   a sailing ship with three or more masts, with fore-and-aft sails on the aftermost mast and square sails on the other masts.

barophile   see **piezophile**.

barque   see **bark**.

bathyal   the ocean floor between the edge of the **continental shelf** and 6,500–13,000 feet (2,000–4,000 m) depth.

bathymetry   the seafloor equivalent of topography.

bathypelagic   the zone of the open ocean lying between 2,300–3,300 feet (700–1,000 m) and 6,500–13,000 feet (2,000–4,000 m) depth.

bathyscaphe   a free-diving, self-propelled submersible vessel used for oceanographic research. It has a crew cabin suspended beneath a float filled with gasoline.

bathysphere   an unpowered, spherical diving vessel that is lowered into the ocean on a cable from a winch on the deck of its support ship.

bends (caisson disease; decompression sickness)   a variety of symptoms, usually including dull but often severe pain in the joints, experienced by divers who surface too quickly. It is caused by the rapid reduction in pressure, which allows nitrogen and other inert gases that high pressure has forced into solution in the blood and tissues to bubble out of solution.

benthic   inhabiting the seafloor.

benthosphere   a submersible designed by Otis Barton that resembled the **bathysphere** but was capable of descending to 10,000 feet (3.050 m).

bioluminescence   the production by a living organism of light without heat.

black smoker   a **hydrothermal vent** discharging water that is black because of its high content of iron, manganese, and copper.

blob (globster)   an unidentified mass of organic matter found on a shore. Blobs are often mistaken for sea monsters.

brig   a fast, maneuverable sailing ship with two square-rigged masts.

caisson disease   see **bends**.

cephalopod   a member of the class (Cephalopoda) of mollusks that includes the living squids, octopuses, cuttlefish, and nautilus, and the extinct ammonites and belemnites.

conformal projection   a map projection that represents all angles and shapes correctly.

conic section   a plane figure that is formed when a plane intersects a cone. If the plane is perpendicular to the conic axis, the resulting figure is a circle; if the plane is inclined with respect to the axis, the figure is an **ellipse**; if the plane is parallel to the sloping side of the cone, the figure is a parabola; the figure is a hyperbola if the plane cuts through both halves of a cone formed from two cones joined at their points on the same axis.

continental drift   the movement of continents in relation to each other.

continental shelf   the section of seafloor that slopes gently from the coast to a depth of about 500 feet (150 m).

corvette   a small, lightly armed, maneuverable warship, similar to a **sloop of war**.

cryophile   see **psychrophile**.

Curie temperature   the temperature above which a magnetic material loses its alignment with any external magnetic field, because its atoms vibrate so vigorously they break free from their mechanical coupling.

dead reckoning   a method of estimating the position of a ship or aircraft by using the heading and time spent on that heading to estimate the distance covered since its last positively determined position.

decompression sickness   see **bends**.

deep sound channel (SOFAR channel)   a horizontal layer of the ocean, from near the surface in latitudes higher than 60° to about 0.6 mile (1 km) below the surface in the Tropics, where the speed of sound is at a minimum. Low frequency sound waves can travel through this layer for long distances without dissipating.

diatom   a single-celled aquatic organism that has a cell wall made from silica and consisting of two halves. Most **phytoplankton** are diatoms, of which there are more than 10,000 species.

diatomaceous ooze   a soft sediment found on the deep ocean floor, consisting of more than 30 percent by volume of the silica shells of single-celled algae called **diatoms**.

drill string   the set of pipes that extend from the collar on board a drilling ship or platform to the cutting bit at the tip of the drill.

echo sounding   measuring the depth of the ocean by emitting a sound signal vertically downward, measuring the time that elapses between the emission and the arrival of the reflection (echo), and converting the time lapse into a distance using the speed of sound in water.

ellipse   a **conic section** formed by a plane that is at an angle to the axis of the cone. It has a long major axis and short minor axis, both passing through the center of the ellipse, and two foci.

ellipsoid (spheroid)   the solid figure that is formed when an **ellipse** is rotated about its major axis.

epipelagic   the region of the open ocean lying below the **oceanic** zone and extending to a depth of about 656 feet (200 m).

equal-area projection (equivalent projection)   a map projection in which areas of surface features are represented accurately.

equidistant projection   a map projection that shows distances correctly, but only between certain points and in certain directions.

equivalent projection   see **equal-area projection**.

extremophile   an organism that thrives under extreme conditions of high or low temperature, acidity, alkalinity, salinity, pressure, radiation, aridity, or other environments usually considered inhospitable.

fathometer   an echo sounder (see **echo sounding**) used to measure sea depth or as a fish finder, to locate shoals of fish.

fault   a rock fracture caused when stresses on either side of a weakness exceed the strength of the rock and it breaks, the two sections moving in relation to each other.

fish finder   see **fathometer**.

frigate   a warship that is smaller, faster, and more lightly armed than a battleship; in the days of sail a frigate was a three-masted vessel, square-rigged on all masts.

globster   see **blob**.

Gondwana   a **supercontinent** comprising South America, Africa, Madagascar, India, Sri Lanka, Australia, New Zealand, and Antarctica, that formed the southern section of **Pangaea**.

graben   a region where the Earth's crust has subsided between two sections of crust that have moved apart.

gravimeter   an instrument used for measuring gravitational acceleration.

great circle   a line on the surface of a sphere that is an arc of a circle centered at the center of the sphere.

guyot   a flat-topped **seamount**.

hadal   the deepest part of the ocean floor, below about 20,000 feet (6,000 m) depth.

hadalpelagic   the zone of the open ocean lying between 20,000 feet (6,000 m) and the deep ocean floor at about 33,000 feet (10,000 m) depth.

halophile   an **extremophile** that thrives in saline environments, requiring at least 2M (2 moles/liter) of salt (NaCl).

heterocercal tail   a fish tail in which the vertebrae extend into the tail, turning upward, producing an upper section of the tail that is markedly larger than the lower section.

hydrographer   a person who measures oceans, seas, lakes, and rivers and plots information about them onto charts.

hydrophone   a microphone designed for use below the surface of water.

hydrostatic equation   the equation used to determine the pressure exerted by water at any given depth below the surface. The equation is: $p = g\rho z$, where p is pressure; g is acceleration due to gravity (32.175 feet per second per second [9.807 m/s^2]); ρ is the density of water (for seawater, 64 pounds per cubic foot [1.027 tonnes {t}/m^3]); and z is the depth below the surface. This is the pressure exerted by the water;

unless the water is sealed from the atmosphere, the atmospheric pressure acting on the water surface must be added; at an air temperature of 59°F (15°C) this averages 14.7 pounds per square inch (101.325 kilopascals [kPa]).

hydrothermal vent a place on the ocean floor on or adjacent to a mid-ocean ridge or ocean trench, where water emerges that has been heated by contact with hot rock beneath the floor, commonly to about 570°F (300°C).

hyperthermophile an **extremophile** that thrives in temperatures of 176°F–250°F (80°C–121°C) and fails to reproduce at temperatures lower than 194°F (90°C).

ichthyologist a zoologist who specializes in the study of fishes.

ichthyology the scientific study of fishes.

illicium the modified fin ray, with a fleshy lure at the tip, possessed by anglerfishes; the fish uses it to lure prey.

Laurasia a **supercontinent** comprising North America and Eurasia that formed the northern section of **Pangaea**.

lithosphere the rocks of the Earth's crust together with the brittle, uppermost layer of the **mantle**.

littoral pertaining to the shore between the limits of high and low tide.

living fossil member of a living species that is almost identical to a species otherwise known only from the fossil record, showing that it has evolved only very slowly, changing little over many millions of years.

loxodrome see **rhumb line.**

luciferase an enzyme that catalyzes the oxidation of **luciferin**, causing the emission of light.

luciferin five basic types of organic compounds, each found in a different group of organisms, that emit energy in the form of visible light.

lye a strongly alkaline solution, commonly used for washing and making soap, that is obtained by soaking wood or seaweed ash in water; originally the active ingredient was potassium carbonate, but in modern use the term includes sodium hydroxide and potassium hydroxide.

magma hot rock that results from the partial melting of the base of the Earth's crust and upper **mantle**.

mantle the region of the Earth lying between the crust and the core; it is approximately 1,430 miles (2,300 km) thick.

master (sailing master) a sailor, not necessarily a commissioned officer, who is in charge of a ship's steering and navigation.

meridian a line of longitude.

mesopelagic the zone of the open ocean lying between 656 feet (200 m) and 2,300–3,300 feet (700–1,000 m) depth.

midshipman in the U.S. Navy, an officer cadet; in the British Royal Navy, the lowest officer rank.

neritic the uppermost layer of the ocean, lying above the **continental shelf**.

oceanic the uppermost zone of the open ocean.

osmophile an **extremophile** that thrives in environments with a very high sugar concentration.

Pangaea a **supercontinent** comprising all of the Earth's lands that formed about 265 million years ago and broke apart into the supercontinents of **Gondwana** and **Laurasia** about 175 million years ago.

Pannotia a **supercontinent** that existed during the Neoproterozoic era.

Panthalassa (Panthalassic Ocean) an ocean that surrounded the **supercontinent** of **Pangaea**.

pelagic inhabiting open water.

petard a small explosive device often employed to force open doors.

pharyngeal jaws a second set of jaws located in the pharynx, behind the mouth, in many fish species.

photic the zone of the ocean, extending from the surface to a depth of about 650 feet (200 m), that is illuminated by sunlight.

photophore an organ that contains colonies of bioluminescent bacteria; see **bioluminescence.**

photosynthesis the process by which green plants synthesize carbohydrates from atmospheric carbon dioxide and water absorbed from the soil, using light energy to drive the reactions.

phytoplankton members of the **plankton** that perform **photosynthesis**.

piezophile (barophile) an **extremophile** that thrives under pressures of 130 tons per square inch (38 MPa) or more.

plankton very small aquatic organisms that drift with the tides and currents in the upper layers of the sea.

plate tectonics the unifying theory of the Earth sciences, which proposes that the Earth's rigid crust is broken into a number of blocks, called plates, that move in relation to each other.

projection any technique that is used to represent the features of a spherical surface on a plane surface.

psychrophile (cryophile) an **extremophile** that thrives in temperatures lower than 59°F (15°C).

radiometric dating a technique for determining the age of a specimen by measuring the relative proportions of its content of a radioactive element and the stable end-product to which it has decayed at a precisely known rate.

rebreather a device used in spacecraft, submersibles, and wherever humans must survive for prolonged periods isolated from the atmosphere. The rebreather passes exhaled air through **soda lime** to remove carbon dioxide and through calcium chloride to remove excess moisture.

receiving hulk a naval ship, stripped of its guns and masts, that is moored, usually in a river, and used for storage.

reflection seismology (seismic reflection) a technique for exploring below the surface of the Earth's crust and the structure of ocean basins by emitting a source of seismic waves and studying their reflections.

refraction the change in direction of a wave as it crosses the boundary between two media with different densities.

rhumb line (loxodrome) a line on the surface of the Earth that intersects all **meridians** at the same angle, consequently tracing a curved path.

Rodinia a **supercontinent** that existed during the Mesoproterozoic era.

routier see **rutter**.

rutter (routier or ruttier) a set of written directions for navigating at sea; the word is from the French *routier*.

ruttier see **rutter**.

sailing master see **master**.

seamount a mountain on the ocean floor that is not high enough to emerge above the surface.

seismically active a region of volcanic and earthquake activity.

seismic reflection see **reflection seismology**.

seismic wave a shock wave generated by rock movement within the Earth's crust.

seismograph a record of **seismic waves** as a set of lines on a graph.

seismology the scientific study of earthquakes and the transmission of shock waves through the Earth's crust.

seismometer an instrument that detects **seismic waves**.

side-scan sonar a technique for mapping features on the ocean floor by emitting a sound signal horizontally from a submerged transmitter, measuring the time that elapses between the emission and the arrival of the reflected sound, and converting the time lapse into a distance using the known speed of sound in water.

sloop of war small three-masted warship with a single gun deck carrying up to 18 cannon.

soda lime mixture of about 75 percent calcium hydroxide (Ca[OH]$_2$), 20 percent water (H$_2$O), 3 percent sodium hydroxide (NaOH), and 1 percent potassium hydroxide (KOH), which is used to remove carbon dioxide (CO$_2$) from air.

SOFAR channel *so*und *f*ixing *a*nd *r*anging channel. See **deep sound channel**.

sonar a type of **echo sounding** used to detect objects underwater; the name is an acronym for *so*und *na*vigation *r*anging.

spar a stout pole to which a sail is attached on a sailing ship.

spheroid see **ellipsoid**.

strike-slip fault a **fault** in which the blocks are displaced horizontally.

subduction the process in which a section of the Earth's crust is pulled beneath an adjacent section and into the **mantle.**

subduction zone a region in which **subduction** is occurring.

supercontinent a continent that forms when several continents move together and become joined.

tectonic plate one of the sections into which the Earth's crust is divided.

Tethys a sea that filled a deep indentation on the eastern side of **Pangaea** and an ocean that extended later from Central America to Southeast Asia.

thermophile an **extremeophile** that thrives at temperatures of 140°F–176°F (60°C–80°C).

transform fault a **fault** that occurs in the rocks of the ocean floor on a boundary between **tectonic plates**, in which the displaced sections move in the opposite direction (their motion is transformed) to the movement in a similar fault occurring on land.

upwelling   the rise of water from the ocean floor all the way to the surface that occurs in particular regions of the oceans.

vellum   fine parchment, originally made from the skin of a calf, used for writing manuscripts and drawing maps; modern vellum is a type of paper that imitates true vellum.

white smoker   a **hydrothermal vent** that discharges water that is white because of its high content of arsenic and zinc.

zooplankton   members of the **plankton** that do not perform **photosynthesis**, instead deriving nutrients by feeding on members of the **phytoplankton**.

FURTHER RESOURCES

BOOKS AND ARTICLES

Banister, Keith, and Andrew Campbell, eds. *The Encyclopedia of Aquatic Life.* New York: Facts On File, 1985. A comprehensive description of marine life.

Bright, Michael. *There Are Giants in the Sea: Monsters and Mysteries of the Depths Explored.* London: Robson Books, 1989. An account of many stories of sea monsters, with descriptions of those that are real and explanations for many that are illusory, by a senior producer with the BBC Natural History Unit.

Brink-Roby, H. "*Siren canora*: The Mermaid and the Mythical in Late Nineteenth-Century Science." *Archives of Natural History* 35 (2008): 1–14. A scholarly account of the debate about mermaids, which ranged from the pages of *Punch* to learned societies.

Gould, Charles. *Mythical Monsters: Fact or Fiction?* London: W.H. Allen, 1886. Reissued in facsimile form, London: Studio Books, 1992. A 19th-century discussion of sea serpents and other mythological animals by an eminent naturalist.

Hancock, Paul L., and Brian Skinner, eds. *The Oxford Companion to the Earth.* New York: Oxford University Press, 2000. A reference book containing explanations of many of the central concepts of the Earth sciences.

Hutchinson, Stephen, and Lawrence E. Hawkins. *Oceans: A Visual Guide.* London: Reader's Digest, 2004. A comprehensive and lavishly illustrated account of the oceans, including their exploration, exploitation, and marine life.

Kunzig, Robert. *The Restless Sea: Exploring the World beneath the Waves.* New York: W.W. Norton, 1999. A journalist's account of the exploration of the ocean floor and deep-ocean life.

Nybakken, James W., and Mark D. Bertness. *Marine Biology: An Ecological Approach.* 6th ed. New York: HarperCollins, 2004. A standard textbook on marine life.

Ogilvie, Brian W. *The Science of Describing: Natural History in Renaissance Europe.* Chicago: Chicago University Press, 2006. A scholarly account of the development of ways of describing plants and animals in the 15th and 16th centuries.

Payne, Ann. *Medieval Beasts.* London: British Library, 1990. A richly illustrated description of medieval animals, real and mythical.

WEB PAGES

Academy of Achievement. "Robert D. Ballard, Ph.D." Available online. URL: http://www.achievement.org/autodoc/page/bal0bio-1. Accessed July 11, 2008. Brief biography of Robert Ballard.

Argo. University of California at San Diego. Available online. URL: http://www.argo.ucsd.edu/index.html. Accessed July 24, 2008. Argo home page, linking to detailed information about every aspect of the Argo program.

Beebe, William. "A Dark and Luminous Blue." *Half a Mile Down,* 1934. Ocean Planet: Writings and Images of the Sea. Washington, D.C.: Smithsonian Institution. Available online. URL: http://seawifs.gsfc.nasa.gov/OCEAN_PLANET/HTML/ocean_planet_book_beebe1.html. Accessed June 30, 2008. Beebe's own description of his 1934 record bathysphere dive.

Bossard, David C. "Report of the Scientific Results of the Voyage of HMS *Challenger* during the Years 1873–76." Hanover, N.H.: Dartmouth College. Available online. URL: http://19thcenturyscience.org/HMSC/HMSC-INDEX/index-linked.htm. Accessed May 2, 2008. Links to the original text of the reports of the expedition.

CIESM: The Mediterranean Science Commission. Available online. URL: http://www.ciesm.org. Accessed July 25, 2008. Home page of the CIESM.

Hebrew University of Jerusalem Institute of Chemistry. "Reginald Aubrey Fessenden." Available online. URL: http://chem.ch.huji.ac.il/history/fessenden.html. Accessed May 8, 2008. A biography of Fessenden, listing his many inventions.

Heidorn, Keith C. Weather People and History. "Admiral Robert Fitz-Roy: The Rest of the Story." The Weather Doctor, 2006. Available online. URL: www.islandnet.com/~see/weather/history/fitzroy-rest.htm. Accessed April 25, 2008. A brief biography of FitzRoy.

Hughes, Patrick. "The Meteorologist Who Started a Revolution." *Weatherwise* 47 (April 1994): 29. Available online. URL: www.pangaea.org/wegener.htm. Accessed June 3, 2008. An outline of Wegener's theory of continental drift.

Integrated Ocean Drilling Program. Available online. URL: http://www.iodp.org/index.php. Accessed July 24, 2008. IODP home page, with links to every aspect of the program and its results.

International Council for the Exploration of the Seas. ICES home page. Available online. URL: http://www.ices.dk/indexfla.asp. Accessed July 25, 2008. Full information about the ICES and its work.

Lewis, Cherry L. E. "Rock Stars. Arthur Holmes: An Ingenious Scientist." *GSA Today* (March 2002): 16–17. Geological Society of America. Available online. URL: http://gsahist.org/gsat/gt02mar17_16.pdf. Downloaded June 4, 2008. A biography of Arthur Holmes.

McCartney, Mark. "William Thomson: King of Victorian Physics." Institute of Physics. December 1, 2002. Available online. URL: http://physicsworld.com/cws/article/print/16484. Accessed June 6, 2008. A biography of William Thomson (Lord Kelvin).

Natural History Museum. "The HMS Challenger Expedition, 1872–1876." London: Natural History Museum. Available online. URL: www.nhm.ac.uk/resources-rx/files/44feat_challenger_expid_1872-3119.pdf. Downloaded October 17, 2008. A brief account of the expedition.

O'Connor, J. J., and E. F. Robertson. "Gerardus Mercator." St. Andrews, U.K.: University of St. Andrews. Available online. URL: www-history.mcs.st-andrews.ac.uk/Biographies/Mercator_Gerardus.html. Accessed April 17, 2008. A brief biography of Mercator.

———. "Regnier Gemma Frisius." St. Andrews, U.K.: University of St. Andrews. Available online. URL: www-history.mcs.st-andrews.ac.uk/Biographies/Gemma_Frisius.html. Accessed April 17, 2008. A brief biography of Gemma Frisius.

———. "William Thomson (Lord Kelvin)". St. Andrews, U.K.: University of St. Andrews. Available online. URL: www-history.mcs.st-andrews.ac.uk/Biographies/Thomson.html. Accessed June 10, 2008. A brief biography of Lord Kelvin.

Pirsson, Louis V. *Biographical Memoir of James Dwight Dana (1813–1895).* Washington, D.C.: National Academy of Sciences, December 1919. Available online. URL: http://books.nap.edu/html/biomems/jdana.pdf. Downloaded May 1, 2008. A fairly detailed biography of Dana.

Smithsonian Institution National Museum of Natural History. "The Coelacanth: More Living than Fossil." Available online. URL: http://www.mnh.si.edu/highlight/coelacanth/index.htm. Accessed July 23, 2008. Description of the coelacanth and its discovery.

State University of New York at Stony Brook. "Hot Vents." Available online. URL: http://life.bio.sunysb.edu/marinebio/hotvent.html. Accessed July 11, 2008. A description of life around hydrothermal vents with photographs of vents and vent organisms.

United Nations Educational, Scientific, and Cultural Organization. "Intergovernmental Oceanographic Commission." Available online.

URL: http://ioc-unesco.org. Accessed July 25, 2008. Home page of the IOC.

University of Oxford. "The Measurers: A Flemish Image of Mathematics in the Sixteen Century." Oxford: Museum of the History of Science. Available online. URL: www.mhs.ox.ac.uk/measurer/text/mathemat.htm. Accessed April 17, 2008. An account of the mathematicians, map makers, and globe makers who were working in Antwerp and Louvain in the 16th century.

Watson, J. M. "Harry Hammond Hess: Spreading the Seafloor." USGS. Available online. URL: http://pubs.usgs.gov/gip/dynamic/HHH.html. Accessed May 14, 2008. A brief biography of Hess and a description of his work.

Wertenbaker, William. "Rock Stars. William Maurice Ewing: Pioneer Explorer of the Ocean Floor and Architect of Lamont." *GSA Today* (October 2000): 28–29. Geological Society of America. Available online. URL: http://gsahist.org/gsat/gt00oct28_29.pdf. Downloaded May 13, 2008. Biographical and career details about Ewing.

Wildlifeonline. "The Megamouth Shark *(Megachasma pelagios)*." Available online. URL: http://www.wildlifeonline.me.uk/megamouth.html. Last updated June 12, 2006. Accessed July 21, 2008. A description of the megamouth shark and accounts of sightings of it.

Wilkinson, Jerry. "History of the Gulf Stream." General Keys History. Available online. URL: http://www.keyshistory.org/gulfstream.html. Accessed June 19, 2008. An account of the discovery and exploration of the Gulf Stream.

Wolff, Torben. "Galathea Report: Scientific Results of the Danish Deep-Sea Expedition Round the World 1950–52." Galathea Committee, Zoological Museum, Natural History Museum of Denmark, 1957. Available online. URL: http://www.zmuc.dk/inverweb/Galathea/index.html. Accessed October 17, 2008. Twenty-volume scientific report on results of the Galathea 2 expedition.

INDEX

Note: *Italic* page numbers indicate illustrations; page numbers followed by *m* indicate maps; page numbers followed by *t* indicate tables.

A

abyss. *See* seafloor *entries*
abyssalpelagic zone *50*
abyssal plain 45
acidophile 142
acorn worms 134
active sonar 122
Adelaide (queen of Great Britain) 23
adenosine triphosphate-adenosine diphosphate reaction 136
ADS (atmospheric diving suit) 111–112
Adventure, HMS 19, 23
Aegean Sea 49
AFB-14 (U.S. Navy research vessel) 166
age, of Earth 81
The Age of the Earth (Holmes) 81
Agulhas Current 102
air, in diving bells 105
air pressure 87–88, 106
air supply, for bathyscaphe 117
air tanks 117, 124
Alexander VI (pope) 86
algae 127
alkaliphile 67–68, 143
Allonautilus scrobiculatus 174
Alps 40, 72
Aluminaut (submersible) 122–124
Alvin (submersible) 123–126, *125*, 138, 143
American Revolution 99

Ampère, André-Marie 94
Amphitrite 147
Amundsen, Roald 20
anaconda 170
Anastasia Island, Florida 160
A new systeme of geography (Seller and Seller) 7
anglerfishes 136, 137
Anglia Contracta 6–7
Angola 160
angular momentum, conservation of 41–42
animals, deep-sea xiii, 140
Antarctica 19
Appalachian Mountains 38
Aqua-Lung 71
Archaea 142, 143
Architeuthis 162, *164*
Arcturus (ship) 114
Argo floats 176–177
Aristotle 13
Arrogant, HMS 27
arsenic 140
Arthropoda 174
aspidochelon 154–155
Assyrian mythology 147
asthenosphere 60, 69
Atargatis 147
Atlantic Current 100–102
Atlantic Ocean 37
Atlantis II (support ship) *125*, 126
Atlas Anglicanus 6
Atlas Maritimus, or the Sea-Atlas: Being a Book of Maritime Charts (Seller) 5, *6*
atmospheric diving suit (ADS) 111–112
atmospheric pressure 87–88
Auguste Piccard (mesoscaph) 121
Australia 87

azimuthal projection 10, 11, *11*
azoic hypothesis 48–51, 128

B

B-52 bomber 123
bacteria 67, 140–142, 184
Bacteria (classification) 142, 143
Baillie sounding sampler 54, *55*
Balaenoptera musculus 168
Balard, Antoine-Jérôme 93–96
Ballard, Robert 138–139, 142–144
Banks, Sir Joseph 18
barbel 135, 136
Barometric Pressure: Researches in Experimental Psychology (Bert) 109
barophile. *See* piezophile
Barton, Frederick Otis, Jr. 112–114
basins. *See* ocean basins
basking shark 128, 167–168
bathymetry 61–62
Bathynomas giganteus 134
bathypelagic zone *50*
bathyscaphe xiii, 115–121, *116*, *118*, *120*. See also *Trieste*, USS
bathysphere xiii, 112–114, *113*
Beacon, HMS 49
Beagle, HMS 23, 24, *24*, 26–27, 28*m*
beard worm 67
Beaufort, Francis 23–24, 26
Beaumont, Jean-Baptiste-Armand-Louis-Léonce-Élie de 36
Beauve, Pierre de Rémy de 106–107
Becquerel, Antoine-Henri 35–36, 79
Beebe, William 112–115, *113*
Behaim, Martin 9
Behm, Alexander 56

bends (caisson disease; decompression sickness) 108–110
Ben Franklin (submersible) 143
Bengal 102
benthosphere 113–114
Bergen sea serpent 155–156
Bergman, Torbern Olof 93
Bert, Paul 109–111
bestiary 150
Bible 151–153
biological classification systems 142
bioluminescence 113, 135–136
black dragonfish 135–136
black seadevils 137
black smoker *139*
    Alvin and 124, 126
    emissions from 140
    Galápagos Rift 138
    Mariana Trench 67
blob (globster) 160–161
blue whale 168
Boa constrictor 170
Board of Trade (England) 28
Bonnycastle, Charles 53–54, 56
Botany Bay 18
bottom-trawl net 132
Bowditch, Nathaniel 31, 32
Boyle, Robert 109
Brabant, Holland 9
Brandywine (frigate) 31–32
breathing 104, 105
breathing apparatus 105–108, 112
breathing tube 105
Brekhovskikh, Leonid Maksimovich 64
Brethren of the Common Life (*Broeders des gemeenen levens*) 13
British Association for the Advancement of Science 149
British Isles *22m*
bromine 95, 96
Brunswick (Norwegian tanker) 162, 164

Bruton, Mike 175
Bruun, Anton Frederik 130–131, 133
bubbles 109
Buchanan, John Young 45
Bullard, Sir Edward 82

C
caisson disease. *See* bends
calcium chloride 112
calcium hydroxide 112
Calyptogena magnifica 140
Cambrian 81
canyons. *See* submarine canyons
Cape Horn 32
Cape Johnson, USS 69, 132
Cape of Good Hope 20
carbohydrates 127
carbon dioxide 127, 140
Carcinoscorpius sp. 174
Carpenter, William Benjamin 48, 51, 128, 130
Carroll, Lewis 2
Carta marina (Map of the sea) 155
Catalina Island, California 167
catalysis 136
cephalopod 161, 174
Cetacea 152
Cetorhinus maximus 128, 167–168
Challenger, HMS xiii, 44–52, *47*
    and Challenger Deep 65
    and Mid-Atlantic Ridge 58–60
    and sea serpents 149
    soundings taken by 54
Challenger Deep 47, 65–68, *66m*, 119–121
Chanticleer, HMS 24
Charles I (king of England) 8
Charles II (king of England) 8, 21
Charles V (Holy Roman Emperor) 13
charts 4
Charybdis 151–152

Chauliodus 135
chemical analysis 46–47, 93–95, *95t*
Chesapeake Bay serpent ("Chessie") 158–159
Chikyu (Japanese research vessel) 180
chimney 140
Christianity 153. *See also* Bible
Christian VIII (king of Denmark) 130
chronometer 12, 19, 100–101
Church of England 8
CITES (Convention on International Trade in Endangered Species) 175
Civil War (U.S.) 33
classification, biological 142
Clerke, Charles 20
climate, ocean currents and 184
clock. *See* chronometer
coelacanth 170–175, *172*
Coelacanth Conservation Council 175
Coelacanthiformes 174
Coleridge, Samuel Taylor 30–31
Colladon, Jean-Daniel 52, 53
colossal squid 164
Cols, Herbert 160
Comoro Islands 173
compass 90
composite organisms 149–150
Confederate navy 33
conformal projection 10
Congo Canyon 63–64
conical projection 10, 11, *11*
conservation of angular momentum 41–42
Consort (brig) 32
continental drift 72–78, 81–84
contraction of Earth 35–37, 41–42
Convention on International Trade in Endangered Species (CITES) 175

Cook, James 16–21, 18*m*
cooling, of Earth 35–38
core, of Earth 35–36, 42
Coretter, Dunham 160
corselet 108
Courtney-Latimer, Marjorie
171–173
Courtois, Bernard 94
Cousteau, Jacques-Yves 71, 123
Cox & Stevens, Inc. 112
Cremer, Gheert. *See* Mercator,
Gerardus
crinoid 51, *51*
Crocodylus niloticus 153
crust, of Earth 42, 84
crustaceans 136, 137
cryophile. *See* psychrophile
Curie temperature 82
currents. *See* ocean currents
Currents of the Atlantic Ocean
(Rennell) 100
CURV (Cable-controlled
Underwater Research Vessel)
123
CUSS 1 (oil-drilling ship) 178
cylindrical projection 10, 11, *11*

D
Daedalus, HMS 156–157
Dana, James Dwight 37–40
Danish Expedition Foundation
131
Darwin, Charles 25–27, 29, 149,
174
Darwin, Erasmus 26
Darwin, Sir George Howard 42–43
Das Antlitz der Erde (Suess) 73
Davy, Sir Humphry 94
dead reckoning 8, 101
decomposers 134
decompression sickness. *See* bends
decompression tables 110
deep-sea drilling 177–180
Deep Sea Drilling Project (DSDP)
178–179

deep-sea submersibles xiii, 69,
122–126, *125*, 177. *See* also
specific submersibles, e.g.: Jago
deep sound channel (SOFAR
channel) 64
Deep Submergence Vessel (DSV)
123
Delaware, USS 38
density, of seawater 46–47, 106
Dénys de Montfort, Pierre 160
*De piscium et aquatilium
animantum natura* (On the
nature of fish and aquatic
animals) (Gessner) 151
Depot of Charts and Instruments
32
depth, of ocean 65, 85–93, 106.
See also soundings, taking
The Depths of the Sea (Thomson)
129
Descartes, René 34–35
diatom 127
diatomaceous ooze 120
*Die Entsehung der Kontinente
und Ozeane* (The origin of
the continents and oceans)
(Wegener) 76
Dietz, Robert Sinclair 71–72
Discovery, HMS 20, 21
*A Dissertation on Elective
Attractions* (Bergman) 93
distortion 10–11
Dive to the Edge of Creation (film)
138
diving 104, 108–110
diving bell 105–107
diving helmet 107–108
diving suits xiii, 106–108, *107*,
111–112
doldrums 30–31
dolphins 148
donkey engine 129
dragonfish 135–136
dredge 128–130
drilling, deep-sea 177–180

drill strings 178
DSDP (Deep Sea Drilling Project)
178–179
DSV (Deep Submergence Vessel)
123
Dunaliella salina 143
Du Toit, Alexander Logie 79

E
Eagle, HMS 16
Earth, contraction of 35–37,
41–42
earthquake 62–63
East India Company 102
*Eccentricities of the Animal
Creation* (Timbs) 148
echo sounding 52–54, 56, 65
economic exclusion zones (EEZ)
181
Edison, Thomas Alva 57
eelworms 134
Egede, Hans 156
Elapidae 169
Elcano, Juan Sebastián 89
Electric Boat Division, General
Dynamics 122
ellipsoid (spheroid) 10
El Niño *87*, 87–89
Emden (German ship) 131
Enchuyser zeecaertboeck
(Enkhuizen sea-chart book)
(Waghenaer) 5
Endeavour, HMS 17–19
English Civil War 8
ENSO (El Niño–Southern
Oscillation) 88
Enteroctopus dofleini 161
Epinephelus itajara 155
epipelagic zone *50*
equal-area projection (equivalent
projection) 10
equidistant projection 10
equivalent projection. *See* equal-
area projection
Erdapfel (earth apple) 9

erosion 37–38, 40
Eukarya 142, 143
Eunectes sp. 170
Everest, Mount 45
evolution 26, 39–40, 149
Ewing, William Maurice 63–65
Exocoetidae 151
extinction 174
extremophiles 142–143
eyelight fish 136

F

The Face of the Earth (Suess) 73
Falmouth (sloop of war) 32
false unicorn 152
Family Herald: A Domestic Magazine of Useful Information and Amusement 149
FAMOUS. *See* French-American-Mid-Ocean Undersea Study
fastitocalon 154
fathometer 56
fault 61–62
Fessenden, Reginald Aubrey 56–58
The First Attempt of the Natural History of Norway (Pontoppidan) 156
Fisher, Osmond 42–43
Fisher, William 5, 7
fish finder 56
fishing 3–5
FitzRoy, Charles 21
FitzRoy, Robert 21–29
flashlight fishes 136
flow, of ocean. *See* ocean currents
flying fish (Exocoetidae) 151
flying fish (serra) 150–151
FNRS-2 115
Folger, Timothy 97, 98
foraminiferans 130
Forbes, Edward 48–51, 128
fossils 72–73
France 99–100

Franklin, Benjamin 96–100
Franklin, James 99
Freminet, Sieur 107
French-American-Mid-Ocean Undersea Study (FAMOUS) 124, 138–139
Frew, Robert 158
Fricke, Hans 173
Fungi 143
Furneaux, Tobias 19

G

Gagnan, Émile 71
Galápagos Rift 124, 138
Galathea expeditions 130–133
Galatheocaris abyssalis 132
Ganges, HMS 23
gas mask 111
gasoline 116–117
Gastropoda 135
Gay-Lussac, Joseph-Louis 94
Gemma Frisius, Regnier 9, 12, 13
General Dynamics Electric Boat Division 122
Geneva, Lake 53
A Geological Comparison of South America with South Africa (Du Toit) 79
geologic timescale 77*t*–78*t*, 81
geology 26
George III (king of Great Britain) 21
Gessner, Conrad 150, 151
giant squid 159–164, *164*
giant tubeworm *141*
Gigantocypris agassizi 134–135
globe 9, 12
globster. *See* blob
Glomar Challenger 178–180
Glossopteris flora 73
goliath grouper 155
gombessa 173
gondola 115–119, 121
Gondwana 41, 73, 73*m*
Goosen, Henrik 170–171

Gore, John 20–21
graben 42
Grassle, J. Frederick 138, 139
Gratiolet, Louis-Pierre 110
gravimeter 68–69
gravitation 35, 68–69
Gray, Asa 39
great circles 36
Great Eastern, SS 58
Greece, ancient 104–105
Greek mythology 147
Green, Charles 17
Greenland 78, 156
Greenwich Hospital 19
Grey family 165–166
Grønningsoeter, Arne 162
grouper 155
Gulf of Aden 82
Gulf of Mexico 53–54
Gulf Stream 97–99, 102
gulper eel 135
guyots 69

H

hadalpelagic zone *50*
hadal zone 133
Haldane, J. B. S. 110, 111
Haldane, John Scott 109–111
half-life (for nitrogen decompression) 110
Halley, Edmond 105
Halomonas salaria 143
halophile 143
Harrison, John 19, 101
hatchetfish 136–137
Hawaiian Islands 20, 166
heat, of Earth's core 42
Heemstra, Phil 173
Henslow, John Stevens 25
herbivores 127
Hercules 151–152
Hess, Harry Hammond 69–71, 82
heterocercal tail 167
Heyden, Gaspard van der 13

Histoire naturelle générale et particulière des mollusques (Dénys de Montfort) 160
Historia de gentibus septenrionalibus (Magnus) 155
Historia naturae maxime peregrinae (Natural history, most especially the foreign) (Nieremberg) 150
historical maps 6, *14*
History of Ocean Basins (Hess) 69
History of the Northern People (Magnus) 155
Holmes, Arthur 79–82
Homer 104–105
horseshoe crabs 174
hot springs 143
Houston, Sam 31
Hudson Canyon 64
The Hunting of the Snark: An Agony, in Eight Fits 2
hydra 151
hydrogen bombs 123
hydrogen sulfide 141
Hydrophis semperi 168
Hydrophis spiralis 170
hydrophone 56
hydrostatic equation 106
hydrothermal vent 67, 138–143, *139, 141. See also* black smoker; white smoker
hyperthermophile 143
"Hypothesis of a Liquid Condition of the Earth's Interior Considered in Connexion with Professor Darwin's Theory of the Genesis of the Moon" (Fisher) 43

I

Iberian (British steamer) 158
icebergs 56
ICES. *See* International Council for the Exploration of the Seas
Idiacanthus 135

Iliad 104–105
illicium 135
Imperial Commissioner for Immigration (Mexico) 33
Indian Ocean 82
Indonesia 87, 88
Integrated Ocean Drilling Program (IODP) 179–180
International Council for the Exploration of the Seas (ICES) 181, 183
iodine 93–94
iodine chloride 95
IODP (Integrated Ocean Drilling Program) 179–180
Iolaus 151
iron 82
iron monosulfide 140

J

Jago (submersible) 174
James II (king of England) 8
Jarrett, Jim 111–112
Java Trench 132
Jeffreys, Sir Harold 43
Jenyns, Leonard 25
JIM 1 diving suit 111–112
Job (biblical book) 153
JOIDES (Joint Oceanographic Institutions for Deep Earth Sampling) 178, *182*
JOIDES Resolution (research vessel) 180, *182*
Jonah 153, 154
Journal and Remarks (Darwin) 26
Judaism 153

K

Kaikō 66, 67
Karoo, South Africa 80*m*
KC-135 refueling plane 123
Kelvin, Lord 89–93
Kelvin and James White Ltd 91
Kelvin White sounding machine 91–92

King, James 21
King, Phillip Parker 23
king crabs 174
Klingert, Karl Heinrich 107
Knorr (research vessel) 143
krait 168, *169*
kraken 159, 161
krill 167, 168

L

Lalla Rookh (sailing yacht) 90
Lamont-Doherty Earth Observatory (LDEO) 65
land bridges 76
La Niña 88
La Pression barométrique: Recherches de physiologie expérimentale (Bert) 109
lateral compression 37–38, 40
Laticauda sp. 168, *169*
Latimeria chalumnae 172, 173–175
Latimeria menadoensis 173, 174
latitude 101
Laurasia 73, 73*m*
Lavoisier, Antoine-Laurent 93
layers, of ocean *50*
LDEO (Lamont-Doherty Earth Observatory) 65
Lenche, Henning M. 132
Leopold (king of the Belgians) 33
Leslie, Forbes 149
Leuven, University of 12, 13, 15
Leuven School 12, 13, 15
leviathan 152–155
Liebig, Justus von 95
light 127
Lightning, HMS 51
Limulus polyphemus 174
Linophrynidae 137
lithosphere 69
living fossil 132–133, 173, 174
London, England 7, 148
longitude 12, 100–102
Löwig, Carl Jacob 95

loxodrome. *See* rhumb line
luciferase 136
luciferin 136
Lulu (support ship) 124
lures 136
Lusitania, RMS 111–112
Luther, martin 13
lye 94
Lyell, Charles 25, 26, 149

M

Magellan, Ferdinand 85–86,
　88–90
magma 69
magnetic field, Earth's 82–83
Magnus, Olaus (Olaf Månsson)
　155
mail packets 97, 98
manganese nodules 185
mantle 60, 81
mapping of oceans 1–33
　Cook's journeys 16–21
　FitzRoy's South American
　　voyages 23–26
　map projection 8–11, *11*
　Maury's mapping of currents
　　29–32
　Mercator's journey 8–9, 12–15
　Seller's atlas 5–8, *6*
　Waghenaer's "mirror" 4
"A Mapp of New England" (Seller)
　5
Mariana Islands 65, 89
Mariana Trench 65–68, 119
The Mariner's Compass 7
Mariner's Mirror 4
mass 41
mathematics 10
Matthew (biblical book) 151
Matthews, Drummond Hoyle 82,
　83
Matthews, Richard 23, 26
Maury, Matthew Fontaine 29–33,
　60, 99
McCormick, Robert 27

McKinley, Mount 45
median trench 137, 138
Mediterranean Sea 72
Megachasma pelagios 166–167,
　167
megamouth shark 166–167, *167*
Melanocetus 137
Melville, Herman 153
Mercator, Gerardus 8–9, 12–15
Mercator projection 9, 14, 15
meridian 101
mermaid 147–150
mermen 147–150
Mesonychoteuthis hamiltoni 164
mesopelagic zone *50*
mesoscaph 121
Meteor Deep 61
Meteorological Office (England)
　28
meteorology 21, 32–33, 74–75,
　176
methane hydrates 185
microorganisms 184
Mid-Atlantic Ridge 59m
　Alvin and 124
　discovery of 58–62
　DSDP 179
　hydrothermal vent on *139*
　seismic reflection for study
　　of 63
　sonar image *61*
Mid-Atlantic Rise 60
Middle Ages 8–9, 150–151
mid-ocean ridges 60, 137
Mielche, Hakon 130, 131, 133
missing link 149
Moby-Dick (Melville) 153
modern exploration 176–183
Mohorovicic discontinuity 177–
　178
Mola mola 165–166, *166*
mollusks 135
monkfish 152
Monodon monoceros 152
Monoplacophora 133

monster anglerfish 137
monsters, real/mythical. *See* sea
　monsters
Moon, origin of 42–43
Morgan, William Jason 83–84
morss piscis 150
mountain chains 36–37
Mozambique Channel 173
M'Quhoe, Peter 156–157
Murray, John 45
mussels 140
mussel shrimp 134
mythology 147, 151–152

N

Nares, George Strong 45
*Narrative of the Surveying
　Voyages of His Majesty's Ships
　Adventure and Beagle* (FitzRoy)
　25–26
narwhal 152
National Geographic 138
National Science Foundation 178
Natoliae Sive Asia Minor 14
Naturalis historia (Natural
　history) (Pliny the Elder) 154
naturalists 150
Nature (journal) 43, 145–146
Nautiloidea 174
Nautilus 174
Nautiolidea 174
Naval Electronics Laboratory (San
　Diego, California) 119
navigation xii
　early forms of 1–2
　Kelvin's studies 90
　and longitude 100–102
　Maury and 31, 32
Navy, U.S. 29, 119, 123, 143
Nematoda 134
Neopilina galatheae 133
Neptune (yacht) 102
Neptunism 37
Nerine (trawler) 170
neritic zone *50*

Netherlands 3–5
*New American Practical
    Navigator* (Bowditch) 31
New Iarsey (New Jersey) 6*m*
*A New Theoretical and Practical
    Treatise on Navigation* (Maury)
    32
New York Zoological Society 114
New Zealand 18, 19, 27, 164, 174
Nieremberg, Juan Eusebio 150
Nile crocodile 153
nitrogen 109, 110
nodules 184–185
Norse mariners 1, 2
North Atlantic Ocean 37
North Holland School 4
North Pacific giant octopus 161
Northumberland, HMS 17
Northwest Passage 19–20
Norway 155–156
nutrition 140–141

O

oarfish 164, *165*
ocean basins xii–xiii, 35–43
ocean currents xii, 29–32, 30*m*,
    96–102
Ocean Drilling Program (ODP) 179
ocean floor xii–xiii. *See also*
    trenches
    Challenger expeditions 45–52
    charting of 52–54
    and continental drift 72–78,
        81–84
    cross-sectional diagram *46*
    Dietz's explorations 71–72
    echo sounding 56
    Mid-Atlantic Ridge 58–62
    and seabed sediment 134
    seismic reflection for mapping
        of 62–64
    soundings of 54–56
    studies of 44–84
    Vening Meinesz's studies of
        gravity on 68–69

oceanic trenches. *See* trenches
ocean sunfish 165–166, *166*
octopus 151, 161. *See also* giant
    squid
Octopus giganteus 160
Odobenus rosmarus 150
ODP (Ocean Drilling Program) 179
On the Origin of Species (Darwin)
    27, 149, 174
"On the Physical Cause of the
    Ocean Basins" (Fisher) 43
*On the Principles of Astronomy
    and Cosmography* (Gemma
    Frisius) 12
optical illusions, sea monsters and
    145–146
organic wastes 134
origin of oceans 34–43
    contraction of Earth 35–37
    Dana's theories on basins
        37–38
    Fisher's Pacific Basin theories
        42–43
    Suess's theories 40–41
*The Origin of the Continents and
    Oceans* (Wegener) 76
osmophile 143
Ostracoda 134
otter board 132
otters 145
Otway, Sir Robert Waller 23
Our Wandering Continents (Du
    Toit) 79
oxygen 109, 112, 169

P

Pacific Basin 42–43
Pacific Ocean 87
Palomares, Spain air collision 123
Pangaea 74, 75*m*
Pannotia 74
Panthalassa (Panthalassic Ocean)
    74
passive sonar 122
Peacock (sloop of war) 39

pelagic sea snake 169
pelagic zone *50*
Pelamis platurus 169
Pembroke, HMS 16–17
pendulum 100–101
Pennington, Kathryn 158
Peress, Joseph Salim 111
Peru Current 87, 89
petard 53
pharyngeal jaws 135
Philippine Trench 131–133
Phillips, Craig 158–159
Phillips, Lodner D. 111
*Philosophical Transactions of the
    Royal Society* 17
Photoblepharon palpebratus 136
photophores 135–137
photosynthesis 127
Physeter catodon 153, 161
Physica curiosa (Physical
    curiosities) (Schott) 152
Physical Geography of the Sea
    (Maury) 33, 99
Physiologus 154
physiology 108–111
phytoplankton 127, 133
Piccard, Auguste 115–119, *116,*
    121
Piccard, Jacques 71, 119–121
Piccard, Jean-Félix 117–118
piezophile (barophile) 143
plane chart 9
plane projection 10–11
plankton 127, 136
plate tectonics 76, 84. *See also*
    tectonic plate
Plexiglas 119, 120
Pliny the Elder 154
Plutonism 36–37
Pogonophora 67
polar ice cap 78
polarity, of Earth's magnetic field
    82–83
Polynesia 2
Ponce de Leon, Juan 98

Pontoppidan, Erik Ludvigsen 156, 159
Porcupine, HMS 51, 128–130
Porcupine Bank 128
porpoises 159
Portugal 3, 4, 86
Poseidon 147
postal service 96–97
potassium hydroxide 112
potassium nitrate 94
predators 135–137
pressure. *See* water pressure
Prévost, Louis-Constant 36, 37
Principles of Geology (Lyell) 25, 26
Principles of Physical Geology (Holmes) 81
Pringle, E. H. 145–146
projection 8–11, *11*, 14, 15
Project Mohole 177–178
Project Nekton 119
protein 141
Protestant Reformation 13–15
Psalms (biblical book) 153
psychrophile (cryophile) 143
Puerto Rico Trench 61
Punch magazine 149
Python reticulates 170

R
radar 56
radio 57–58
radioactive decay 42, 79–81
radiometric dating 80
radius 41
Rangoon, SS 146
rat-trap fishes 136
rebreather 116
receiving hulk 48
Red Sea 137–138
reflection seismology 62–64
Reformation 13
refraction 63, 148
Regalecus glesne 164, *165*
religion 26
Rennell, James 100–103
Rennell Island 168

Rennell's Current 102
Report on the Mollusca and Radiata of the Aegean Sea (Forbes) 49
Resolution, HMS 19, 20
reticulated python 170
Reynolds, Julian Louis 122
Reynolds Marine Services 123
Reynolds Metal Company 122
Rhincodon typus 167–168
Rhizocrinus lofotensis 51, *51*
rhumb line (loxodrome) 14
ridges 84
Riftia pachyptila 140, *141*
rifts 42
rift valley 42, 60
Rime of the Ancient Mariner (Coleridge) 30–31
robots 176
rock 36–37
Rodinia 74
Romanche Trench 60–61
Rome, ancient xiii, 154
Rosebud (vent site) 140
Rose Garden (hydrothermal vent) 138, 140
rotational radius 41
routier 3–4
Royal Danish Academy of Sciences and Letters 130
Royal Geographical Society 26
Royal Navy 21–23
Royal Society of London 17, 19, 21, 26, 28, 128
Rutherford, Ernest 80
ruttier. *See* routier

S
Saccopharynx 135
Sagami Bay, Japan 167
Sailing Directions (Maury) 30
saltpeter 94
San Antonio (carrack) 86
Sanderson, William 16
Sandwich Islands 19, 20
Santa Maria (freighter) 119
Santiago (carrack) 86

Sars, George Ossian 51
Sars, Michael 51
Satan 153
sawfish (serra) 150–151
scaly dragonfishes 136
Scandinavia 159
Scandinavian mythology 147
scavengers 134
Schott, Caspar 152
Scripps Institution of Oceanography 71, 138, 178–179
scurvy 19
Scylla 152
seabed sediment 46, 54, 134
seafloor exploration 104–126, 128–133
  atmospheric diving suit for 111–112
  bathyscaphe 115–121
  bathysphere 112–114
  and the bends 108–110
  deep-sea submersibles 122–126
  diving bells 105–107
  diving suits 106–108
  Galathea expeditions 131–133
  Porcupine expedition 128–130
seafloor life 127–144
seafloor spreading 60, 69, 71, 76, 83
sea krait 168, *169*
sea level 98
sea lily 51
sea monk 152
sea monsters, real/mythical 145–175
  coelacanth 170–175, *172*
  giant squid 159–164, *164*
  leviathan 152–155
  megamouth shark 166–167, *167*
  mermaid 147–150
  mythical monsters 150–162
  oarfish 164, *165*
  ocean sunfish 165–166, *166*
  sea serpents 155–159
  sea snakes 168–170

seamounts 69
sea serpents 146, 155–159
sea snakes 168–170
seawater, chemical analysis of
    93–96, 95t
seawater pressure. *See* water
    pressure
seaweed 94, 127, 146, 157, 159
Sedco/BP 471 180
sediment. *See* seabed sediment
sedimentary rock 38
seed shrimp 134
seismically active ridges 62
seismic reflection 62–64
seismic wave 63
seismograph 63
seismology 63
seismometer 63
Seller, John 5–8
serra 150–151
Seven Years' War 16–17, 102
shock waves 62–63
shrinking, of Earth's crust 35–37
side-scan sonar 52, 122
Siebe, August 107
Silliman, Benjamin 38, 39
Silurian 174
silver chromate 92
Skyring, W. G. 23
Smeaton, John 105–106
Smith, James Leonard Brierley
    171–173, 175
Smithsonian Institution 161
snorkelers 105
soda lime 112
sodium hydroxide 112
Solander, Daniel Carlsson 18
solar eclipse 17
Solomon Islands 168
sonar 60, 61, 69, 122
sound, speed of 53, 64
sounding line *91*
soundings, taking
    and Challenger Deep 65
    Challenger expedition 46–47
    early methods 54–56
    Kelvin's methods 90–92

by Magellan 85–86
    by Sturm and Colladon 52–53
sound waves 52
South Africa 170–171, 173
South America 23–26, 87, 88
South Atlantic Ocean 37, 179
South Georgia island 19
South Pacific Ocean 162
South Sandwich Islands 19
South Sandwich Trench 61
Spain 3, 86, 88
Speculum Regale (The king's
    mirror) 147
sperm whale 153, 161
Sphenodon sp. 174
spheroid. *See* ellipsoid
Spieghel der zeevaerdt (Mariner's
    mirror) (Waghenaer) 4
spy hopping 148
squid, giant 152, 159–164, *164*
stalked crinoid 51
steamships, taking soundings
    from 90–92
Steenstrup, Japetus 152
Sternoptychidae 136–137
Stokes, Pringle 23, 25
Stomiatidae 136
Strait of Magellan 88
Strait of Malacca 170
strike-slip faults 62
Sturm, Jacques-Charles-François
    52–53
subduction zone 70
submarine canyons 63–64, 71
submarine mountain chains 60
submersibles. *See* deep-sea
    submersibles
subterranean reservoir 34
Suess, Eduard 40–41, 72–73
sulfides 140–141
Sun 43, 101
Sunda Double Trench 132
"Sundry Marine Observations"
    (Franklin) 97, 100
sunfish. *See* ocean sunfish
sunlight 127
supercontinent 38

surface air pressure 87–88
Suroît (French research ship) 143
surveying 8–9
swabs 129–130
swim-bladders 149
swordfish 151
*System of Mineralogy and
    Crystallography* (Dana) 38–39

T
Tachypleus tridentatus 174
Tahiti 17–18, 87
Tardigrada 134
Taylor, Clyde and Carol 158
Taylor, W. H. 111
tectonic plate 83m, 84, 137
telegraphy 89. *See also*
    transatlantic cable
temperature
    Challenger measuring of
        seawater 46–47
    of sea in Gulf Stream 98
    and speed of sound in water
        53
Tethys Sea 41, 72–74
thermophile 68, 143
Thetis, HMS 22
Thompson, William 49
Thomson, Charles Wyville
    Challenger expedition 45,
        47–49, 51–52
    Galathea expeditions 131, 133
    Porcupine expedition 128, 130
Thomson, William (1st baron
    Kelvin). *See* Kelvin, Lord
Thornton, John 5, 7
Thresher, USS 121
Thresoor der zeevaerdt (Treasure
    of Navigation) (Waghenaer) 5
tide gauge 90
tides 42
Tierra de Fuego 23
Timbs, John 148
Titanic, RMS 56, 143
trade 3, 15–16
trade winds 88, 98
transatlantic cable 32, 58, 89

transform fault *61*, 62, *62*, 83
trenches *70*
    and *Galathea* expeditions
        131–132
    and Hess's mid-ocean ridge
        theory 70
    Mariana 65–68, 119
    and Mid-Atlantic Ridge 60–61
    in Red Sea 137, 138
    and seafloor spreading 68–70
Trieste, USS 66, 71, 115–121, *116*,
   118, *120*
Trinidad (carrack) 85–86, 89
Triton 147
tuatara 174
tubeworm 139, 140, *141*
"Twelve Bolt Helmet" 107–108

U

U-28 (German submarine) 158
undersea life 127–144
United Nations Conference on the
   Law of the Sea 181
United States (frigate) 38
Upcott, England 100, 102
upwelling 134

V

Vædderen (Danish surveillance
   vessel) 133
Vann, John T. *113*
vellum 4
Vening Meinesz, Felix Andries
   68–69
vent clam 140
vent organisms 140
Venus 17
Verrill, Addison Emery 160
Victoria (carrack) 89
Vikings 1
Vincennes (sloop of war) 29, 32, 39
Vine, Allyn C. 123
Vine, Frederick John 82, 83

viperfish 135
Virginia Military Institute 33
viruses 184
visibility 105
volcanoes 35, 70
The Voyage of the Beagle (Darwin)
   26

W

W. H. D. C. Wright (merchant
   bark) 29–30
Waghenaer, Lucas Janszoon 4–5
Wales 165–166
Walker, John and Henry 16
walruses 148, 150
Walsh, Donald 119–122
Wandank, USS 119
Washington, USS 53
water bears 134
water pressure
   adaptation of deep-sea
      creatures to 133
   and the bends 109
   and Challenger Deep 67
   and diving 105
   and Kelvin White sounding
      machine 92
   and ocean depth 106
weather 87, 88, 98
Weather Bureau, U.S. 57
weather forecasting map 22*m*
weather satellites 176
Webb, DeWitt 160
Wedgwood, Josiah 26
Wegener, Alfred 73–76, 78, 79
wells 34
Westinghouse, George 57
whales 148, 153–154, 168
whale shark 167–168
White, James 91
white smoker 67, 140
WHOI. *See* Woods Hole
   Oceanographic Institution

Wild, Jean-Jacques 45, 46
Wilhelm (duke of Jülich-Cleves-
   Berg) 15
Wilkes, Charles 39
Willemoes-Suhn, Rudolf von 45
William III (king of England) 8
William IV (king of Great Britain)
   23
Wilson, John Tuzo 82, 83
wind 98
Wind and Current Charts
   (Maury) 30
Woese, Carl 142
Woods Hole Oceanographic
   Institution (WHOI)
      Alvin and 123
      Robert Ballard and 143, 144
      Challenger Deep exploration/
         charting 66
      William Maurice Ewing and
         64–65
      examination of hydrothermal
         vents at Galápagos Rift 138
World War I 56, 76, 111
World War II 64–65, 69, 122
Wrottesley, John 28

X

Xiphias gladius 151

Y

Yale University 38, 39
yeasts 143
yellowbelly sea snake 169
yellow sea snake 170
Yellowstone National Park 143

Z

Zeus 151–152
zinc 140
zones, of ocean *50*
zooplankton 127
Zug, George R. 158